W9-BTI-180

DARK
REMEDY

DARK REMEDY

The Impact of Thalidomide and Its Revival as a Vital Medicine

Trent Stephens and
Rock Brynner

PERSEUS PUBLISHING
Cambridge, Massachusetts

Grateful acknowledgment is made for permission to quote from the following: Samuel Beckett, *Waiting for Godot* (copyright © 1954), granted by Grove Press; Harold Evans and The *Sunday Times* Insight Team, *Suffer the Children: The Story of Thalidomide* (copyright © 1979), granted by Penguin Putnam, Inc.; Sylvia Plath, "Thalidomide," from *Ariel* (copyright © 1963), granted by HarperCollins; Henning Sjöström and Robert Nilsson, *Thalidomide and the Power of the Drug Companies* (copyright © 1972), granted by Penguin UK.

Printed in the United States of America.

Cataloging-in-Publication Data is available from the Library of Congress
ISBN 0-7382-0404-8

Perseus Publishing is a member of the Perseus Books Group.

Find us on the World Wide Web at http://www.perseuspublishing.com

Text design by Jeff Williams
Set in 10.5-point Goudy by Perseus Publishing Services

First printing, January 2001

1 2 3 4 5 6 7 8 9 10—03 02 01

CONTENTS

Do not despair, one of the thieves was saved;
Do not presume, one of the thieves was damned.

—St. Augustine, *The City of God*

PREFACE

BY ROCK BRYNNER

As an historian, I read the proposal for Trent Stephens's book about thalidomide with complete captivation. After further research of my own, the contours of this unique medical detective story began to emerge, embedded in a rich historical panorama, with a cast of remarkable characters who came vividly to life. Retracing its path, I have sometimes felt as if I were walking down a dark corridor with a flashlight and finding, left and right, an unlikely succession of fascinating worlds.

Most of what is known about the early history of this drug in Europe comes from the research of two scientific researchers for thalidomide victims in Sweden, Dr. Robert Nilsson and attorney Henning Sjöström; they later published their findings as *Thalidomide and the Power of the Drug Companies* (1972). Equally, a research group of the *Sunday Times* of London, called the Insight Team, collected hundreds of thousands of documents—many from the drug's manufacturers—that were later condensed in their account of the ordeal, *Suffer the Children: The Story of Thalidomide* (1979). As editor of the newspaper at the time, Harold Evans played a vital role in the European proceedings; we are grateful to him for allowing us to cite extensively from their work.

I am especially grateful to Amanda Cook, senior editor of Perseus Books, and to Dr. Trent Stephens, who together invited me to help tell this story.

PROLOGUE

BY TRENT STEPHENS

I was just beginning high school in the tiny farming town of Malta, Idaho, when the story of thalidomide appeared in the U.S. press: I recall a drawing in *Scientific American* that showed a child with foreshortened arms. During the 1960s dozens of laboratories were anxiously engaged in trying to solve the mystery of how thalidomide caused more than 10,000 malformations worldwide—without success. After high school I attended Brigham Young University, aiming to become a biologist and to discover answers to some of the great mysteries of life, and not thinking much about thalidomide. One of my ambitions was to map the human genome.

But during a lecture in my sophomore year, as Dr. Duane Jeffery discussed the genetics of a complex little marine plant, a single question launched me onto the path I have followed ever since: How is the shape of any biological form determined during development? A paper by Dr. Edgar Zwilling further inspired me to use the developing limb to unlock the mysteries of how biological form unfolds—the study of morphogenesis.

As I became committed to a lifelong study of limb development, thalidomide became intertwined with my professional life. To earn my Ph.D. I joined Dr. Jay Lash at the University of Pennsylvania because of work he had published on thalidomide's mechanism in causing birth defects. I then joined Dr. Tom Shepard, one of the world's leading specialists in birth defects, at the University of Washington. While there, during the late 1970s, I joined the search for thalidomide's mechanism by experimentally evaluating the leading hypothesis, only to find it didn't hold water.

By the 1980s no one knew what more to do to solve the thalidomide mystery, and very few people even cared anymore. In reply to a grant proposal I wrote during that period, a reviewer commented, "Why bother to discover thalidomide's mechanism? The drug is now only of historic value." The drug that had come onto the world scene in the 1950s, killed or maimed thousands of infants, and then disappeared nearly as quickly, was all but forgotten.

But thalidomide has returned, and is now being used to treat everything from arthritis to cancer. Interest in the drug and its mechanism in causing birth defects, and how that mechanism might be related to treating a wide range of ills, has grown steadily during the last decade. So has our basic knowledge of embryonic development, to the point where we can now begin to make some educated guesses as to how the drug works.

This book recounts the saga, starting with the drug's shadowy beginnings, and culminating with the search for causes and cures. The issues raised by thalidomide hold great relevance for us today, from the safety of medicines we take to fight a cold to those that will be invented tomorrow to fight cancer. The solution to thalidomide's many mysteries will open doors onto some of tomorrow's greatest medical achievements.

An Uncertain Utopia

1

Your sheep, that were wont to be so meek . . .
swallow down the very men themselves.
—Thomas More, *Utopia* (1516)

The birth of Utopia in the 1950s proclaimed an era of new dreams. Optimism and energy were everywhere in 1957, as World War II receded into memory and the postwar baby boom reached its crest. "Utopian" was the byword in advertising that year, providing a vision shared by millions thanks to television; by then, one out of ten families already had color TVs, many with remote "Channel Commanders." Science offered a wide array of innovations for everyday life, along with the illusion that Tomorrowland had already arrived. Housewives could serve TV dinners in twenty minutes. At shoe stores, kids spent hours pressed up against X-ray machines, watching the bones of their toes wiggle. And it seemed as if DDT was on the verge of eradicating insects once and for all.

In a series of articles unashamedly entitled "Man's New World," the editors of *Life* magazine featured jet-engine cars, plastic houses shaped like mushrooms, and foam-rubber furniture. The intrepid futurists predicted that "personalized flying machines . . . will open up hitherto inaccessible rural lands for daily communication," and anticipated that disposable clothes and "dehydrofrozen" meals would be common features of modern life by 1977.

This was the dawn of the jet age. Soon Boeing 707s would be crossing continents and oceans in just a few hours. People could drive from coast to coast in record time on Eisenhower's new turnpike system, in the biggest gas guzzlers ever built or in the new German imports, Volkswagens, which could travel thirty-six miles on a gallon of gas.

There were marvels in every field of human endeavor. Univac, the biggest, fastest computer ever built, had just compiled a Bible concordance, 2,000 pages, in less than 1,300 hours. Anything and everything could be made from plastic. Infectious diseases, it seemed, had been all but eliminated with universal vaccinations for smallpox, polio, and most of all with accessible antibiotics. Tuberculosis, pneumonia, and other scourges were no longer among the top ten killers, thanks to breakthroughs in drug therapy. And the average life span had increased ten years just since 1935.

"The power and the pace of technology [has] . . . brought man to the brink of his greatest technological accomplishment," crowed *Life* magazine. "He [is] . . . now ready to rocket himself into the endless emptiness of outer space." Werner Von Braun, the top U.S. rocket scientist, predicted that "we should be able to send men to the moon and back within twenty-five years." In fact, it took half that time. "Whether we know it or not, whether we like it or not," wrote the editors of *Life* in October 1957, "the daily life of each of us is being changed—and is destined to be changed far more—by events taking place in laboratories and factories across the land."

Such bounding progress would come at a cost, but the price was not yet apparent. The same week that that issue of *Life* magazine reached the newsstands, a new "wonder drug" was introduced on the German market that would play a significant role in shattering the complacent vision of the future that enthralled the world of the 1950s.

Utopia was already filled with anxiety before thalidomide cast its grim specter. Daily life proceeded under the hour-by-hour threat of nuclear annihilation. Nuclear war drills were held weekly in schools across America, to rehearse for the bright flash that might come at *any* moment. The fallout shelter business was booming, and Americans were learning to live with unceasing anxiety. Korea, Vietnam, and the Suez Canal reminded the world that World War II had not been the final conflict.

And on October 4, 1957, three days after thalidomide went on the market in Germany, the USSR launched the first *Sputnik* satellite, its radio beeping incessantly while it circled the planet, as if to remind Americans that we were in second place technologically, well behind our archfoe. That notion, especially unsettling in the nuclear age, was reinforced in December 1957, when a Vanguard rocket designed to launch the first U.S. satellite exploded on liftoff. By then, the Soviets were celebrating *Sputnik II*.

Utopia was shaky on the domestic front as well. Anxiety was high across the country as the shadow of slavery was vividly exposed by the new civil rights movement. U.S. troops were now amassed in front of a high school in Little Rock, Arkansas, where nine black children made history by trying to obtain the same education as white students.

Even the medical world was anxious that fall: a new influenza epidemic, called the Asian flu, was poised to strike Europe and the United States. Drug companies everywhere were working overtime to produce new vaccines, as the war on disease escalated.

In the United States it was only the war veterans who had experienced the horror of bomb attacks, but in England and Germany, no one over the age of twelve could forget their nights of terror and days of grief. English civilians from London to Coventry had endured years of aerial bombing. The V-1 and V-2 rocket attacks had left tens of thousands of men, women, and children shell-shocked and psychologically crippled. The V-1s, or buzz bombs, were especially terrifying: the sound of their rough, choppy engines was followed by an eerie silence after their fuel ran out—and then came the explosion. The V-2s were less noisy, but more deadly.

In Germany the closing years of the war had been living hell. Allied bombers flew overhead at will, night and day, laying waste to major cities and factory towns. When German citizens went to sleep at night, they could never feel sure they would awake in the morning. Then, in 1945, came the ground invasion. Berliners who had fled east to avoid the bombs ran headlong into the advancing Russian army, and reversed their flight back to Berlin.

Little wonder that, twelve years later, many British and Germans found it hard to sleep at night.

Tranquilizers and sleeping pills played a large role in the uncertain Utopia of the 1950s. One doctor testified in Congress that "the people of this nation are being steadily educated by doctors and the drug industry to take a drug whenever they felt anxious about anything." For many people, another testified, drugs were "used as a panacea to solve personal problems." In Great Britain an estimated 1 million people used some type of sedative daily, and about one out of eight National Health Service prescriptions was for sleeping pills. Almost all of the tranquilizers were dangerous barbiturates. Deaths from accidental and deliberate overdose were on the rise; in fact, suicide by sleeping pills was the glamorous way to check out. In 1955, the United States produced almost 4 billion barbiturates, or twenty-six pills for every man, woman, and child in the country. According to Senator Hubert Humphrey, one out of every seven Americans took barbiturates.

The U.S. pharmaceutical industry was now launching over 400 new drugs every year. Prescriptions had nearly quadrupled over the past twenty years, and drug exports had increased twentyfold since World War II. There were pills for everything. Chemists had just announced a drug that could speed up suntanning ("Next summer, something you swallow may turn you the color of a life guard!"); human tests were under way on inmates at Arizona State Prison. The culture was also beginning to learn that some drugs could be very dangerous. In the United States, the "Feds" were beginning to crack down on illegal pep pills like Dexedrine. Long-haul truckers relied on Benzedrine ("bennies" or "co-pilots") to stay awake. There was already a booming black market in these amphetamines at truck stops around the country, and almost anyone could order large quantities of the drug through the mail.

Aldous Huxley was predicting, as he had in both *Brave New World* and *The Doors of Perception* that even though most people still relied upon alcohol to forget Communist threats and society's woes, before long a new pill would be produced to help people unwind. Reconsidering Huxley's fond dream is peculiar today, since we have progressed from "Mother's Little Helper"—by the Rolling Stones—in the sixties to father's little helper—Viagra—in the nineties, and offering a whole smorgasbord of psychotropics, from Prozac to Ecstasy. In June 1956 Huxley wrote an article in the *Sunday Times* of London, observing that *Homo sapiens* had been taking mind-altering drugs since prehistory—especially alcohol, of course. "Will the pharmacologist be able to do better than the brewers and distillers?" Huxley wondered. "It seems reasonable to suppose it."

An executive at a British pharmaceutical firm, Distillers Company (Biochemicals) Ltd., read Huxley's article and promptly pointed it out to the company's director, E. G. Gross, in a memo the next day. "The ultimate target," he wrote, "would be the production of the ideal tranquilizing agent to replace alcohol among those people who would prefer to 'transform their minds' by this alternative means." Gross replied, ". . . it will not be long before there are as many of these things as there are brands of whisky."

The very same week, Chemie Grünenthal offered Distillers Company an opportunity to license thalidomide for manufacture and distribution in the United Kingdom. From the way it was described by the German company that had invented it, this new sedative was the dream drug for the Utopian market that chemical companies around the world were aiming to conquer. It seemed to the Distillers executives like the answer to their prayers; less than a month after the Huxley article, the company brought quantities of thalidomide back from Germany for testing.

Chemie Grünenthal was another hungry pharmaceutical company, though not yet one of the well-established corporations. Grünenthal was a family-owned business formed in 1946 that initially produced ointments, cough medicines, disinfectants, and herbal medicines in an abandoned, seventeenth-century copper foundry (Kupferhof), built like a fortress out of stone in the small West German village of Stolberg, near Aachen. The closest large city, Düsseldorf, has been called "the desktop of the industrial region" for a century, home to the executive bureaus of coal mines, steel plants, and other heavy industry. Most of the city, apart from the Altstadt, had been heavily bombed and not yet rebuilt when Grünenthal began operations nearby in 1946. In those years, Düsseldorf was not an especially cheerful city. Even today the main boulevard, Heinrich Heine Allee, commemorates the local poet well known to doctors for "Morphine," a poem about Heine's own terrible medical ordeal: "*Gut ist der schlaf, Der Tod ist besser, Das beste were nie geboren sein.*" ("Sleep is good, Death is better, The best is never to have been born at all.")

Grünenthal was a subsidiary of a large cosmetics company. Their research was unashamedly market-driven, and their initial corporate strategy was to penetrate the burgeoning antibiotic boom. Conditions in postwar Germany were appalling, and health authorities feared epidemics of tuberculosis and even cholera. So antibiotics were big business for German pharmaceutical companies.

The director of Chemie Grünenthal's research and development group was Dr. Heinrich Mückter, a man of cold, hard judgment who had joined the company in 1946. His arrogant appearance was chilling, not unlike British actor Christopher Lee cast as an evil scientist, his right eyebrow permanently arched in contempt. A strategist, not a caregiver, Dr. Mückter was staunchly, blindly committed to the family business that had taken him in during that terrible year of defeat in Germany, amid growing revelations about what the army had done in places like Poland, and the genocide for which Germany was responsible.

Two years before he joined Grünenthal, Mückter was a medical scientist for the army of the Third Reich. Specifically, he was Medical Officer *(Stabsarzt)* to the Superior Command of the German Occupation Forces occupying Krakau, Poland, with the additional, ominous title of "Director for the Institute of Spotted Fever and Virus Research." The German Army was not renowned for missionary medical work in Poland. Given the role that military medicine played in the objectives and methods of the Nazi occupation of Krakau, one can only presume that Mückter's work there involved the science of killing rather than healing.

Within a year at his first civilian job, Mückter succeeded in producing penicillin from mold cultures, and thereby boosted Grünenthal into the limelight: it became the first company in West Germany to be allowed by the Allied military government to produce penicillin industrially. By 1950 they were producing the drug in both oral and injectable form.

In the race to find new antibiotics, it wasn't long before Grünenthal became notorious for rushing drugs to market with inadequate testing. Between 1953 and 1954, the company developed an antibiotic they called Supracillin, a variant of streptomycin. Though powerful, streptomycin is also highly toxic: the drug may damage nerves between the inner ear and the brain, resulting in deafness. Grünenthal claimed they had completely eliminated these side effects with Supracillin, a claim that later proved untrue. The company also developed another antibiotic, Pulmo 550, which, they claimed, was superior to penicillin. But Grünenthal had ignored reports of severe side effects, and was later harshly reprimanded for using this antibiotic in humans before testing it thoroughly in animals.

The antibiotic market soon became saturated, and Grünenthal branched out into research on other new drugs. Under Mückter's command was Wilhelm Kunz. Kunz had served as a sergeant during the war, and he appears to have sustained a character of dogged obedience into civilian life. Kunz, who had very little background in the sciences, was made the chief of chemical research for Grünenthal's research and development team, engaged primarily in mixing organic compounds to find something with pharmacological potential. And this is key: both Mückter and Kunz were contractually entitled to a percentage of the profits from any new drug they developed.

A third member of the team, pharmacologist Dr. Herbert Keller, was in charge of testing the new pharmaceuticals developed by Kunz and his people, but he too had very little experience. Mückter later acknowledged that the company's department of pharmacology was "without a qualified leader" from June 1957 until February 1960 and that Dr. Keller was "not particularly well known in pharmacological circles."

Sometime in early 1954, under instruction from Mückter, Kunz set to work trying to find a simple, inexpensive method for manufacturing antibiotics from peptides, the bonds that hold amino acids together to form biologically active molecules—proteins. One of the experiments conducted by Kunz in his attempt to make peptides was to heat a commercially available chemical named phthaloylisoglutamine. The molecule thereby produced was α-phthalimidoglutarimide, which they called thalidomide. Its unusual molecular structure was composed of three rings.

As Dr. Robert Brent of DuPont Children's Hospital in Delaware later observed, "It was made from two rather innocuous compounds each of which you could take in large amounts with no effect whatsoever . . . both very safe compounds. When you put them together, we find a very powerful birth defect–causing agent." The same molecule had been produced briefly by a Swiss company, which had found it ineffective in animal tests and abandoned research. But Grünenthal secured a twenty-year patent for the drug and began testing it in animals and humans and cast about for some disease that thalidomide could cure.

A drug in search of a disease: surprising as it may seem, that is not an unusual approach to creating pharmaceuticals. AZT was first developed in 1964, but failed as an anticancer drug; then, in 1987, as every company searched for a drug to defeat AIDS, Burroughs Wellcome Co. tested AZT in patients with the human immunodificiency virus; the first trials showed such dramatic efficacy that the trials were cut short, and within just one week, it was approved by the FDA. But the Grünenthal team soon learned that thalidomide was not an antibiotic or an antihistimine, and it did not appear to affect tumors in mice. It even failed as a sedative in animals; that is, it had no effect on "righting reflexes" in rats or on their treadmill "holding reflexes," two classic tests of sedation in animals.

In fact, the only thing thalidomide seemed to have going for it was that researchers could not find a dose high enough to kill a lab rat. The toxicity test results were remarkable, setting thalidomide apart from almost all other compounds. In these trials, animals are fed a chemical to determine the dosage at which half of the tested animals die; this is called the LD_{50}. As Dr. Keller's group began testing thalidomide, they found that, even at extremely high doses, thalidomide simply did not kill rats. They also tested the drug on mice, guinea pigs, rabbits, cats, and dogs. Since no other side effects were observed in any of the animals tested, they declared the drug nontoxic. Given the shocking number of deaths from barbiturates, the people at Grünenthal reasoned that a nonlethal sedative could capture a large portion of the booming tranquilizer market.

Despite the fact that thalidomide showed no sedative effects in animals, Herbert Keller was intrigued by its structure, which appeared to him similar to that of barbiturates, the most widely-used class of tranquilizers. But this assessment owed more to inexperience than to chemical shrewdness. Thalidomide does not have the structural similarities that Keller imagined, a CIBA research team later proved; in fact, it does not fit into any of the structural classes of sedatives. But a tranquilizer without toxicity would have enormous market potential. So even though thalidomide did not sedate animals, Keller was authorized by Mückter to try it as a sedative in humans. With that, they began human clinical trials.

This was ground zero for history's greatest medical tragedy. Based on the contemporaneous practices of other drug manufacturers, the company should have conducted cautious, investigative human trials with thalidomide—including its impact on pregnant women. Just since 1949, in countries around the world, twenty-five compounds had been found that killed or deformed the fetus, establishing the need for such trials. In 1953, for example, the results of human trials of cortisone in pregnant women were published, noting any malformations in their children at birth. Such a study of thalidomide would have limited the tragedy to a few individuals.

Instead of conducting a methodical investigation, the Grünenthal team simply distributed free samples through doctors, without any monitoring or follow-up, to judge from the little evidence that remains. They even handed out pills among their employees. As one observer put it, "Thalidomide was introduced by the method of Russian roulette. Practically nothing was known about the drug at the time of its marketing." Exactly why the Grünenthal men acted so irresponsibly remains, at best, a subject for speculation; but the damaged medical world of postwar Germany—the "cradle of thalidomide"—suggests that the lingering callousness of medicine under the Third Reich played a role.

Just seven years earlier, twenty-three elite German doctors had been tried in Nuremberg for their crimes against humanity, and sixteen had been found guilty. Known as "The Twenty-Three," these doctors were only tokens for the whole cruel medical machine. In its military medical research and its spurious, demented torture of prisoners in concentration camps in the name of "research"—throughout Nazi-occupied territory, but most notoriously at Auschwitz by Dr. Josef Mengele—the German medical community, including thousands of outstanding caregivers, had operated in a climate governed by genocide and war.

Medical science provided the governing metaphor of Nazi genocide: genetics, after all, is a medical science. A decade before Mückter, Kunz, and Keller began testing thalidomide in humans, a large, influential proportion of German doctors was collectively committed to mass murder. The "euthanasia" program began in hospitals and sanitariums across Germany eight years before the gas chambers went into operation. It was driven by the underlying conceit they shared, with or without zeal: that doctors had to heal the Fatherland by excising the "diseased" parts of society—citizens identified by the chilling phrase "life unworthy of life."

"The entire Nazi regime was built on a biomedical vision that required the kind of racial purification that would progress from sterilization to extensive killing," according to Robert Jay Lifton in his sweeping study, *Nazi Doctors: Medical Killing and the Psychology of Genocide*. The euthanasia program was established in 1939 to eliminate the incurably ill; that is, the undesirable, a category that grew exponentially, and decreed that *every murder had to be performed by a doctor*, according to the dictate that "the syringe belongs in the hand of a physician." But the syringe quickly proved inadequate. At Auschwitz alone:

> Nazi doctors presided over the murder of most of the one million victims of that camp. Doctors supervised the killing in the gas chambers and decided when the victims were dead. Doctors kept up a pretense of medical legitimacy: for deaths of Auschwitz prisoners and of outsiders brought there to be killed, they signed false death certificates listing spurious illnesses. . . . Doctors were given much of the responsibility for the murderous ecology of Auschwitz—the choosing of victims, the carrying through of the physical and psychological mechanics of killing. . . . While doctors by no means ran Auschwitz, they did lend it a perverse medical aura. As one survivor who closely observed the process put the matter, "Auschwitz was like a medical operation. . . . The killing program was led by doctors from beginning to end."

It would be foolish and simplistic to suppose that Grünenthal's slipshod methodology in testing thalidomide was somehow a last ember of Hitler's empire. The connection was not that direct. The decision to proceed so swiftly to negligent human trials of thalidomide was the responsibility of a few individuals, not of the postwar culture. It was probably a decision made in utter ignorance by men who simply did not know what they were supposed to do. But that ignorance was a product of their specific medical education and culture, and the fact remains that Grünenthal's actions and arguments reflected the culture of the Third Reich, not the new Germany of Adenauer and Brandt. German medicine in 1954 was still reemerging from a grotesque, nationwide biomedical fantasy that had collectively distorted all norms for determining what human testing was necessary and appropriate.

It is difficult to reconstruct much of Grünenthal's research, because most of their documentation was destroyed or "misplaced" even before the German court proceedings began in 1968 because, according to Mückter, "it seemed unnecessary to keep the individual test protocols for the future." (He later contradicted himself, claiming that the records had "disappeared during moving of the files some time during 1959.")

But documents were collected early on by a Swedish lawyer, Henning Sjöström, and his scientific colleague Robert Nilsson, who went on to work for and to testify on behalf of Swedish victims. As well, tens of thousands of documents were collected by the "Insight Team" of the *Sunday Times* of London beginning in 1967, and for the next twelve years a half-dozen determined journalists, including the paper's editor, the formidable Harold Evans, compiled documents by the tens of thousands and interviewed many key figures.

"Only in Grünenthal's published table of animal results," the *Sunday Times* team concluded, "can the wondrous phenomenon be perceived of a highly potent drug with no sign of toxicity. In no one else's hands have these phenomena been brought together by experimental data. The conclusion that the Grünenthal men saw what they wanted to see . . . is difficult to avoid."

It is also an unavoidable conclusion that the key executives of the company simply did not care, and decided to ignore the possible consequences, even to their own careers, if thalidomide proved dangerous. Four years after the company began selling the drug over the counter, they received a letter from a doctor in Finland, asking if thalidomide could cross the placenta. Grünenthal's answer was, "Not known." If it did cross the placenta, could it harm the fetus? "Unlikely," came Grünenthal's best guess.

A popular myth, dating back to the Middle Ages and still prevalent in the 1950s, was that nothing harmful could ever cross the placenta from mother to embryo. But it had been known since 1955 that any substance with a molecular weight of less than 1,000 could cross the placenta and enter the fetal blood. The molecular weight of thalidomide is 258.

It is still a matter of debate today whether adequate animal testing in the 1950s would have revealed the teratogenic potential of thalidomide. Many companies then routinely conducted animal tests to determine if new drugs would be safe if taken by pregnant women. In a classic work of teratology, it had been demonstrated in

1948 that the dye known as trypan blue could cause birth defects in rat embryos, whereas the mother rats exhibited no symptoms.

Thalidomide is a most unusual teratogen, however: at routine doses, thalidomide does not cause birth defects in any animals other than primates—which were not used for such tests until years later, largely as a result of the thalidomide epidemic. Senator Estes Kefauver of Tennessee was oversimplifying the problem when he later declared to the Senate that, "if thalidomide had been tested in rabbits . . . this whole catastrophe would have been avoided." Rabbit testing only reveals birth defects in quantities 150 times greater than the therapeutic dose—and such extreme tests were not commonly conducted; in fact, they were only conducted after the thalidomide disaster, when researchers were *trying* to cause birth defects that did not show up in a routine screening. In August 1962 Dr. Helen Taussig wrote in *Scientific American* that, in her laboratory at Johns Hopkins School of Medicine, "I have not been able to obtain abnormalities in baby rabbits with thalidomide primarily because the massive doses I have used bring on so many abortions. . . . If thalidomide had been developed in this country, I am convinced that it would easily have found wide distribution before the terrible power to cause deformity had been apparent."

In the absence of any scientific rationale, early in 1955 Grünenthal began human trials in West Germany and Switzerland. Thalidomide was apparently first tested as an anticonvulsant for epileptics. It did not prevent convulsions, as would be expected of a sedative, but it did cause patients to go into a deep, all-night, "natural" sleep. These early clinical trials demonstrated that thalidomide is not just a sedative that produces relaxation; it is actually a hypnotic, which directly induces sleep. This was better news than anything Grünenthal had hoped for.

As the trials continued, thalidomide was found to have a soothing, calming, and sleep-inducing effect on patients without decreasing motor activity. Some side effects were seen in a few patients. These included shivering, buzzing in the ears, a hangover the next morning, nausea, constipation, or a rashlike allergic reaction. Some patients even experienced giddiness or wakefulness instead of the desired hypnotic, sleep-inducing effects. These symptoms were seen in only a small number of subjects, however, and were considered insignificant. Many of the physicians involved in the trials were impressed by the drug's hypnotic effect.

One problem still had to be solved by Grünenthal: even though the drug was effective in humans, it could not go to market without some confirmation of its effects in animals. (Of course, the correct method of drug testing is to demonstrate an effect in animals *first*, then to test the drug in humans.) So when Keller's group was unable to demonstrate any sedative or hypnotic effects in animals by conventional trials, they devised a decidedly unconventional test.

The team developed what Keller called a "jiggle cage," in which eight mice were suspended over a sulfuric acid bath, like hapless victims in some ridiculous B movie. A platinum wire hung from the bottom of the cage, and the slightest movement of the mice caused the wire to dip into the sulfuric acid. The chemical reaction of the platinum with the acid released hydrogen gas, and the more hydrogen measured, the more active the mice were said to be. Lacking anything that resembled peer review of their criteria, Keller and his group simply declared that a "hypnotic state" was reached in the mice when this sedative reduced the amount of hydrogen released from their device by half, even though the mice were still wide awake.

Using the jiggle cage and their unusual criterion for sedation, Grünenthal was able to "demonstrate" that thalidomide was a more powerful hypnotic than comparable sedatives in their study. It was all they needed to go to market. They claimed to have shown that this new drug was a better hypnotic than anything else available—and that it was completely safe.

Thalidomide was first sold in a drug called Grippex, which contained thalidomide quinine, vitamin C, phenacetin, and aspirin. It was marketed on a limited basis for treating respiratory infections in Hamburg in November 1956.

On October 1, 1957, thalidomide was first released as a sedative under the patent name Contergan. Because of its unambiguous claims of safety, thalidomide was sold over the counter in Germany, and later in other countries as well. The marketing campaign was massive. Grünenthal advertised the new drug in fifty medical journals, sent out 50,000 "therapeutic circulars," and mailed 250,000 personal letters to individual physicians, emphasizing the drug's safety. The main thrust of the campaign was that, unlike other sedatives currently on the market, thalidomide was completely safe. Even a determined suicide could not take enough Contergan to cause death. Furthermore, accidental overdoses by children would be unheard-of with this drug, a claim later substantiated by actual cases that were widely publicized by the company.

The marketing campaign was highly successful, and over the next three years sales grew steadily. By 1961, thalidomide was the best-selling sedative in Germany, with total sales of DM 12.4 million, five times as much as its leading competitor, Doriden. Thalidomide sales soon represented half of Grünenthal's total income, and by this time many of the company executives' personal incomes were tied to its sales. The growth of the company was dramatic: in 1954, when the drug was formulated, Chemie Grünenthal had 420 employees; by 1961 the company employed more than 1,300. This was mostly because of the remarkable success of thalidomide.

Sharing the stage with Chemie Grünenthal in the gathering tragedy was the British corporation, Distillers Co. (Biochemicals) Ltd., of Liverpool, England. As the name suggests, it had been an alcohol distillery. During World War II, Distillers had been asked by the British government to manufacture penicillin, the new wonder drug. After the war the government had offered the Liverpool penicillin factory to Distillers at a bargain. Even before the sedative was introduced in Germany, Grünenthal had been looking to expand into the international market; contracts were signed with the Astra Company in Sweden in March 1957 for Scandinavian distribution, and in July with Distillers for market rights throughout the British Commonwealth.

With the Utopian market very much in mind, Distillers had set out to find someone who had already developed a safe sedative. To avoid the cost of research, development, and testing, the company had decided to buy the British rights to a drug that had already been developed elsewhere. Dr. Walter Kennedy, chief medical adviser for Distillers, visited Grünenthal's headquarters in Stolberg, West Germany, in June 1956—the same week the Aldous Huxley article appeared—and reported back to the company about thalidomide. "If all the details about this are true then it is a most remarkable drug. In short, it is impossible to give a toxic dose."

Apparently Distillers believed that no further testing would be necessary to market thalidomide in the United Kingdom; that was one of the attractive features of the deal. In fact, Grünenthal had stipulated that, contractually, Distillers had to begin marketing the drug within nine months, which effectively forbade meaningful trials before the drug went on sale. At the time Distillers didn't even have a pharmacologist on its staff. Dr. George Somers, who joined the company after it had purchased the British rights to thalidomide, said that had he been there to test the compound at the beginning, he would have "thrown it out of the window."

So without conducting any trials of their own, in April 1958 Distillers began distributing Distaval, their nearly eponymous trade name for thalidomide, throughout Britain. It was advertised as "completely safe," the solution to the "mounting toll of barbiturate deaths." Before long thalidomide was outselling its main competitor three to one. Distillers' market included the British Isles, Canada, Australia, and New Zealand. Eventually thalidomide was distributed in forty-six countries throughout Europe, Scandanavia, Asia, Africa, and the Americas. During the peak year of 1961, 25 percent of thalidomide sales were in these foreign countries.

In the world market, thalidomide was sold separately under at least thirty-seven names. It was also combined with a number of other drugs, including aspirin, quinine, and even barbiturates. In several countries it was available by prescription only, but in Germany and other countries it was sold over the counter. Alone or in combination, thalidomide was used for colds, coughs, flu, asthma, headaches, anxiety, and, of course, sleeplessness. Later, when birth defects began to appear, the multiple drug names, the combinations, and the absence of prescriptions made it very difficult to establish the cause of the epidemic. Thalidomide was considered so safe that many women didn't bother to list it among drugs they had taken during pregnancy.

In liquid form the drug was sometimes referred to as "the babysitter," since in Germany many parents gave it to sedate their children in movie theaters; hospitals used the drug to calm children during clinical examinations. In Great Britain, advertisements insisted that accidental ingestion of the drug by children could not be lethal; they displayed a small child taking a bottle from a medicine cabinet with the caption, "This child's life may depend on the safety of Distaval." And, in August 1958, when Grünenthal in a letter to German physicians declared that thalidomide was the best drug for pregnant and nursing mothers, Distillers quickly added the claim to the drug's label in England without any scientific foundation whatsoever. As the *Sunday Times* team put it, Distillers "exhibited a naïve desire to believe only the best of thalidomide," and "accepted everything that Grünenthal said about it with absolute trust, and to promote this apple of its eye . . . was prepared to turn a vacuum of knowledge about thalidomide into specific assurances of safety."

The only major country that had not yet been penetrated was the United States, and Grünenthal executives were eager to crack that enormous market. But when they proposed distributing thalidomide to Smith, Kline and French—a company with much more experience in research—the pharmaceutical giant tested the drug in animals and declined to consider it further. Grünenthal approached several other U.S. pharmaceutical companies without success. Finally, in October 1958, the company signed a contract with the 134-year-old William S. Merrell Company of Cincinnati to market thalidomide throughout the United States (sharing the Canadian market with Horner, Ltd. of Montréal, which sold the drug under the brand name Talimol). On September 8, 1960, Richardson-Merrell (as it was then known) submitted an application to the U.S. Food and Drug Administration to market Kevadon throughout the United States.

The Epidemic 2

The glass cracks across,
The image
Flees and aborts like dropped mercury.

—Sylvia Plath, "Thalidomide" (1962)

On Christmas Day 1956, a baby girl was born without ears.

Her parents lived in Stolberg, West Germany, where Chemie Grünenthal was based. Her father worked for the company, and he had brought home samples of their new drug for his pregnant wife. It was years before he discovered exactly how his daughter had been medically maimed, and learned that she was the first living victim among the many thousands of infant deaths and gravely deformed survivors to come from this avoidable epidemic. It is ironic, of course, that a company employee's child appears to have been the first victim; and it compounds the tragedy, by highlighting how avoidable it was. Had Grünenthal at least tracked the unwitting victims who accepted the medication, the birth of the first little girl without ears—ten months before Contergan went to market—

might have alerted a more diligent scientific team to the public menace that was just beginning. Instead, it was five years until the enormity of the catastrophe could be gauged. The full toll of the catastrophe became incalculable. The full measure of courage summoned up by these thousands of families could never be tallied.

About a year after thalidomide became available to the general public, complaints of side effects began trickling in to Grünenthal. Such user feedback was not uncommon for a new compound, but it was certainly unusual for a drug that was guaranteed to be completely safe. Among other physicians, Dr. Gustav Schmaltz reported to Grünenthal in December 1958 that some of his elderly patients who took thalidomide had experienced giddiness and slight balance disturbances. The company shrugged off these early notices.

During the following year, thalidomide sales increased dramatically. So did the reports of side effects. Complaints included dizziness, decreased blood pressure, hangover, memory loss, constipation, trembling, cold hands and feet that slowly turned numb, and allergic reactions. Many of these symptoms had been seen—and ignored—in the clinical trials.

In October 1959, Dr. Ralf Voss, a Düsseldorf neurologist, wrote to Grünenthal asking if anyone had reported that thalidomide caused polyneuritis, a numbness and tingling in the hands and feet: a sixty-three-year-old patient who had been using thalidomide regularly for a year and a half was exhibiting such symptoms. Drs. Heinrich Mückter and Günther Sievers wrote back that they had received no such complaints. That was a deliberate, bold-faced lie.

Even at this point, when the drug was available over the counter, Mückter and Sievers could have averted thousands of tragedies. A few dozen follow-up calls to their earliest thalidomide patients would have alerted them to a fundamental problem that was not going to disappear if it was ignored. But the executives of Grünenthal were driven by commercial ambition, and repeatedly failed to show any scientific perseverance, much less conscience.

Toward the end of November, Voss wrote to Grünenthal again; this time he had three patients exhibiting the same symptoms, all of whom had taken thalidomide for over a year. Voss believed there was a connection between the long-term use of thalidomide and polyneuritis (also called peripheral neuritis, or neuropathy). Grünenthal replied, "We have no idea how these cases of peripheral neuritis could have been caused by Contergan." Voss was left with the distinct impression that the company believed his claims but did not want his information made public.

Dr. Voss was not deterred when Grünenthal denied there had been other cases of peripheral neuritis. On April 30, 1960, he reported his findings on three patients with polyneuritis at a conference of neurologists in Düsseldorf. Now, as more complaints of nerve damage poured in to Grünenthal, the medical community started hearing about a possible connection between long-term thalidomide use and polyneuritis. At about this time Grünenthal hired a private detective, Ernst Günther Jahnke, to put patients with peripheral neuritis under surveillance, in case they sought compensation; before long, the private eye was also keeping tabs on doctors who criticized thalidomide. In one such case, he reported that the doctor's father was "an ex-communist and nowadays a member of the Social Democrats." The company even sent a woman posing as a polyneuritis patient to one of the drug's critics, but she could report no misconduct on his part.

A new phenomenon began to appear across Germany. Physicians were seeing cases of a birth defect so rare that most had never encountered anything like it. In December 1959, a Dr. Weidenbach presented the case history of a one-year-old girl. The baby's arms and legs were so reduced that the hands and feet were attached directly to the body. This condition is known as tetra-phocomelia (literally, "four seal's limbs.") Weidenbach felt that this case must be genetic. Though not completely unknown, this type of defect was extremely rare; there were only a handful of cases in the literature, and a Danish study in 1949 had estimated that phocomelia occurred in 1 out of 4 million births. Still, a few other physicians had seen similar cases. Just recently, in fact.

Things were not going well for the new wonder drug. Pressure was mounting to put Contergan under prescription. Grünenthal fought this attempt and, suppressing relevant information, never admitted to the cases of polyneuritis that were reported to them; their entire marketing strategy had been built around the guarantee that thalidomide was completely safe. As for the drug going under prescription, Grünenthal's sales manager insisted in an internal memo that "everything must be done to avoid this, since a substantial amount of our volume comes from over-the-counter sales."

On the other side of the world, another story was beginning to unfold. On August 18, 1960, Walter Hodgetts, senior New South Wales sales representative for Distillers Co. Biochemicals (Australia) Ltd., met with Dr. John Newlinds, the medical superintendent at Women's Hospital, Crown Street, Sydney. He was there to persuade the hospital to try Distaval (thalidomide), a sleeping pill that had already been on the market in Britain for two years. Hodgetts left a bottle of Distaval tablets with Newlinds, who in turn sent it on to the hospital pharmacy. During his sales rounds, Walter Hodgetts visited Dr. William McBride, who ran one of the largest obstetrics practices in Australia. McBride agreed to try Distaval as a sedative for his patients in labor, little knowing how this encounter would alter the course of his life, for better and for worse.

Two weeks later a woman came to the Crown Street Hospital emergency room. She had been vomiting for several days, and nothing that anyone tried could stop it. She was two months pregnant, and the physical effort of vomiting was so severe that it threatened to cause a miscarriage. Since all other medications had failed, Dr. McBride decided to try the new drug, and the patient stopped vomiting. McBride was impressed. The medicine appeared to have considerable promise for obstetric patients, and the package insert stated that it was completely safe, even during pregnancy. So McBride began to prescribe Distaval to patients who complained of morning sickness, nervousness, or insomnia.

Although the epidemic of birth defects spread quickly across Germany, few physicians noticed the increase. Almost every pediatric clinic in West Germany had seen cases of phocomelia during the year, but they were not yet aware that others were also seeing cases;

they all believed they were observing isolated events. In October 1960, Drs. W. Kosenow and R. A. Pfeiffer from the Institute of Human Genetics in Münster presented X-rays of two phocomelic babies at a meeting of the German Pediatric Society in Kassel. They presented the cases as interesting, random examples of an extremely rare defect. Dr. Helen Taussig from Johns Hopkins University, who was attending the meetings, said later that "little note was taken of the exhibit. I missed it myself, although I was at the meeting."

How can such an epidemic go undetected? one might wonder at first. But unless a country keeps a registry of all infants born deformed—which is common practice now—one might better ask, how can such an epidemic ever be detected? Often, the answer is: by chance. For example, two women sitting in an ophthamologist's waiting room with their infants, who both had eye problems, began talking, and realized that during pregnancy they'd both had the German measles; it was that coincidence alone that led to the explanation of the teratogenic effect of the rubella virus.

If the German medical community was slow to recognize the underlying cause of the defects, it was in part because of the traditional conviction that a mother's placenta protected the fetus from any toxins, chemical or otherwise. But in fact, research had outdistanced tradition years before; in 1934 H. Marshall Taylor wrote: "That the placenta is permeable to drugs has long been recognized. The obstetrician has for many years accepted this fact." And John B. Thiersch of the University of Seattle later calculated that by 1959 "not less than twenty-five compounds were shown by various investigators . . . to effect the fetus in utero, either killing many fetuses or inducing malformations. . . . " But many German doctors still believed in the "placental barrier," and no one was quick to guess that a drug could cause such specific malformations—especially a "completely safe" drug.

In December 1960 the *British Medical Journal* published the first report of peripheral neuritis in four patients, in a letter from Dr. A. Leslie Florence entitled "Is Thalidomide to Blame?" By that time Grünenthal had received some one hundred reports of severe peripheral neuritis attributable to thalidomide, and fifteen hundred other reports of side effects. In fact, the company had a well-disguised "bunker" beneath an old factory chimney where many incriminating documents were kept. Meanwhile, thalidomide was breaking sales records, in some countries, selling second only to aspirin.

Dr. Voss, still determined to alert the medical community, reported to the medical academy in Düsseldorf in mid-February 1961 that he was not aware of a single case of recovery from peripheral neuritis, even after patients had discontinued the medicine. All the nerve damage caused by thalidomide appeared to be permanent, and in some cases the patients' thumb muscles actually atrophied. In just six weeks, the number of cases reported to Grünenthal had swelled to at least 400. Within Germany, many hospitals had at last discontinued their use of the drug. But the general public had not.

Drs. Fullerton and Kremer, at Middlesex Hospital in London, published an extensive report in 1961 of thirteen cases of peripheral neuritis. Most of the patients were between the ages of fifty and seventy-six. Their typical nightly dose was 100 milligrams of thalidomide; most of the patients had taken the drug for less than a year before the numbness and tingling began, and some had been experiencing these symptoms for as long as two years. Most of the patients had lost sensation in both the feet and hands, and, in the most advanced cases, sensation had been lost throughout the lower limbs up to the groin. One patient had lost sensation in her face. The majority of the patients experienced leg cramps, and some had reduced or absent vibration sensation in the lower limbs. Several of the patients also experienced muscle weakness and wasting in their limbs.

This nerve damage is a pernicious disease that occurs in somewhere between 5 and 20 percent of people who use thalidomide for several months. The nerve damage does not reverse itself after the drug is discontinued; follow-up studies have shown ongoing neuropathy four and even six years later. The damage begins with a prickly feeling in the feet, followed by a slight numbness and coldness in the toes, barely noticeable to the patient. The numbness spreads to the balls of the feet, then to the ankles, and then the calves. In most cases, the loss of sensation does not extend above the knee, but in some cases the entire lower limb may be involved. Days later the fingers begin to feel numb, and then the whole hand. In addition to the numbness there can be severe muscle pain, with cramping in the lower limbs. There is often a loss of reflexes and muscle coordination. Sometimes the patient cannot judge the position of the lower limbs, and walks with an unsteady, ill-coordinated gait. Eventually the patient may experience partial paralysis. At least one person found the nerve damage so unbearable he committed suicide.

These are the symptoms that Chemie Grünenthal chose to ignore, or to discredit as rare, allergic responses. When Dr. Goeden, the company's Cologne rep, visited a clinic, he was asked about the danger of peripheral neuritis and, in his own words, "I did my best to foster confusion."

When Dr. Sievers, from Grünenthal, learned that Dr. Voss planned to publish a paper on the polyneuritis cases he had examined, he visited Voss. But the Düsseldorf neurologist could not be swayed; he had been all but ignored by Grünenthal since reporting the first complaint sixteen months earlier, and he would not be put off now.

By February 1961, the British publications had not had much impact, but after Voss's February 15 publication, everyone at Grünenthal "lost his head." Some of the executives said that thalidomide should be put under prescription immediately, but Heinrich Mückter, co-creator (with Kunz) of thalidomide, wouldn't hear of it. "The fight for Contergan must go on," he declared, "to the bitter end." After all, Grünenthal's company motto was: "We must succeed at any cost."

Mückter argued that the effects of the Voss paper must be neutralized. Although he did agree to place a mild warning on the label, Grünenthal was already launching a campaign through their sales force to create confusion over the issues. The sales people were to deny any connection between Contergan and polyneuritis, and, if denial didn't work, they were to downplay the severity of the disorder and insist that the nerve damage was caused by a rare allergic reaction, suggesting that the whole issue stemmed from rumors sewn by competitors. Under no circumstances was anyone to be informed of the number of cases reported to the company.

Grünenthal did not initiate research in response to these reports. In fact, the company even issued a circular insisting that the polyneuritis might stem from a previously administered sedative or to the misuse of alcohol. To this, one enraged doctor wrote, "You apparently assume that receivers of your circular letter are so unintelligent and uneducated they can be impressed with such simplified propaganda . . . your letter constitutes a monstrous imputation and is unworthy of a serious enterprise." It is clear from this and similar letters that the behavior of Grünenthal's executives was entirely unethical by the standards of the time.

Then, after February 1961, they toned down their "nontoxic" rhetoric in Germany—but not abroad. In forty-five other countries, the campaign of disinformation only grew louder. In a monthly report for April, one executive wrote, "We have managed to bring further delays to unfavorable publications, but after May this will no longer be possible." April marked the peak in Contergan sales for Chemie Grünenthal. Thereafter, sales began to decline because of the bad publicity generated by the peripheral neuritis; that spring, three more papers were published linking thalidomide to nerve damage. By then Grünenthal had sold nearly 64 million thalidomide tablets.

In England, Distillers was just starting to send out leaflets to physicians recommending the use of thalidomide in cases of psychiatry, geriatrics, neurology, dermatology, pediatrics, and obstetrics. Distillers did not mention the mounting problem with polyneuritis, but announced that "Distaval can be given with complete safety to pregnant women and nursing mothers, without adverse effect on mother or child." In fact, no studies of possible birth defects had ever been conducted by Distillers or Grünenthal.

By the end of May, the pressure from physicians to withdraw thalidomide from the market was becoming intense. Contergan was banned in more and more hospitals. A Dr. Paukert wrote to the company, "It is irresponsible to continue marketing this drug," expressing the sentiment of many. But most physicians in Germany were not in the universities or large hospitals, and they didn't keep up on all the medical journals. Because thalidomide was supposed to be nontoxic, they failed to attribute the polyneuritis they were seeing in patients to the drug. So the number of cases actually reported was almost certainly just the tip of the iceberg.

As the reports piled up, Mückter began to recognize that these reports were not hoaxes perpetrated by competitors, but genuine, grave side effects of his medicine. At an internal company meeting on July 14, he actually admitted, "If I were a physician, I would not now prescribe Contergan. Gentlemen, I warn you—I do not want to repeat an earlier judgment—I see great dangers." But he did not alter the company's strategy one whit, and became even more determined to promote to other doctors this drug that he would not himself have prescribed. Six weeks later, at another meeting, Mückter insisted that Contergan was "the best sleeping pill in the world."

Earlier that year a lawyer by the name of Karl Schulte-Hillen went to visit his sister, who had just delivered a baby with extremely reduced limbs that resembled flippers: phocomelia. Six weeks later, Schulte-Hillen's wife delivered a daughter with the same type of malformations. He was convinced that there must be some connection. A gentle man with a powerful build, broad shoulders, and jet-black hair, Schulte-Hillen was driven by conscience and a thunder of emotions to discover the cause.

A few weeks after the birth of his child, Schulte-Hillen met with Professor Widukind Lenz, a pediatrician and head of the children's clinic at Hamburg University, to see if the doctor could offer some explanation, genetic or otherwise. When he showed Lenz the X-rays of his daughter, the pediatrician looked astonished, and left the room. He returned with another X-ray he had just received that morning: it was almost identical to that of Schulte-Hillen's daughter. Unaware that an epidemic of phocomelia was unfolding, Lenz promised to look into the situation. Together, the lawyer and the doctor placed ads in newspapers, searching for children born recently with similar defects.

In Australia, the first of Dr. William McBride's patients treated with thalidomide gave birth; it was the patient who had suffered from severe vomiting at Crown Street Hospital. She delivered a normal, healthy child. But three other babies born soon after were not so lucky. On May 4, McBride attended the birth of "baby Wilson." The baby had radial aplasia (missing radius bones in the forearms) and bowel atresia (blockage), and died a week later. On May 24, McBride attended the birth of another baby with malformations strikingly similar to those of the Wilson baby, and on June 8 came a third baby with similar defects. These two babies also died soon after birth.

At the start of a long weekend holiday, McBride took home the hospital records of the three babies. He began a crash course in teratology—the study of birth defects—about which most physicians knew very little. To his disappointment, McBride didn't find much in the scanty medical literature. One of the items he read, however, was the 1960 CIBA symposium on birth defects and possible environmental contamination; the take-home message from this symposium was that drugs could cause birth defects. That reminded him of an article by John Thiersch, published in 1952, reporting that the drug aminopterin could cause abortion or severe malformations if taken early in a pregnancy. But what McBride was about to discover would eventually move the concept of drug-induced birth defects onto a much more solid theoretical foundation: the hospital records made it clear that the only common factor in the three cases of malformed children was that the mothers had taken Distaval. And he remembered reading a letter to the *British Medical Journal* from Dr. Leslie Florence, suggesting that long-term use of thalidomide might cause nerve damage—so, clearly, it was not an entirely harmless drug, regardless of what Distillers was telling him.

By the end of the long weekend McBride was convinced that Distaval had caused the birth defects. On Tuesday he met with Dr. Newlinds at Crown Street Hospital and suggested as much. Newlinds was quickly convinced by McBride's observations, and that very day he removed thalidomide from the hospital pharmacy. McBride also telephoned Distillers and told a representative there his conclusions. And he began testing thalidomide on mice and guinea pigs.

As a matter of record, Dr. William McBride was the first person to connect thalidomide with birth defects, and that will forever be to his credit. He did so, not thanks to brilliant laboratory research, but in part because, however innocently, he had prescribed the drug to his patients in the first place.

McBride later insisted—and still insists—that on that day, June 14, 1961, he also sent a paper to the *Lancet*, expressing his belief that thalidomide was causing birth defects in Australia. McBride asserts that his paper was returned on July 13 with a cover letter informing him that the *Lancet* had rejected it for publication. The *Lancet*, which generally records submissions, has no record of receiving McBride's paper, and McBride has been unable to produce either his original paper or the *Lancet*'s reply. Had such a paper been published in July 1961, much suffering would have been prevented.

The summer of 1961 was not a good time for Dr. McBride. In the studies he had begun, none of the animals that were fed thalidomide during pregnancy had delivered malformed offspring. Furthermore, there had not been a malformed baby born at Crown Street Hospital since June, even though many mothers who delivered that summer had taken Distaval during pregnancy. McBride began to have doubts. But the hiatus ended in September 1961, when two more thalidomide babies were born at Crown Street. McBride, with confidence in his hypothesis restored, once again called Distillers.

At the end of the summer, despite Grünenthal's efforts to thwart marketing constraints, thalidomide was placed under prescription in Germany, after reports of permanent nerve damage finally came to light, though its greater damage remained unknown. Grünenthal had stalled as long as possible, postponing what was inevitable while building revenues with over-the-counter sales, and causing permanent nerve damage in thousands. The company had lost some market share because of the negative publicity, which, they hoped, would recede now that the drug was under prescription. Surely better times were in store for Grünenthal. That was their thinking on the eve of their worst corporate nightmare.

Widukind Lenz was a thoughtful, good-looking, bespectacled pediatrician with thinning hair and a subdued manner, guided entirely by his concern for patients, not scientific theory. He had a growing sense of dread as the early outlines of a widespread epidemic began to emerge, yet it was still a total mystery. What could possibly account for these deformities? Lenz went to see Dr. Kosenow about cases he had presented with Dr. Pfeiffer almost a year earlier, and learned that Kosenow, together with a geneticist, Professor Degenhardt, had started investigating the outbreak of phocomelia. Lenz offered to study the incidence of birth defects around Hamburg in the preceding two years.

Before computers, that was a heroic undertaking for what was only a hunch, but what he discovered in the statistics was terrifying. He methodically went through all of the city's 212,000 birth records from 1930 to 1955 and found just one case of phocomelia recorded. In the past twelve months, among 6,420 babies born in Hamburg hospitals, there had been eight cases of phocomelia.

What could cause such an outbreak? Then Lenz ran across a paper by Dr. H. R. Wiedemann that described twenty-seven more cases of phocomelia around Kiel, suggesting that one of the newer drugs might be responsible. Wiedemann proposed that the malformations followed a specific pattern that constituted a syndrome. With that, Lenz began open-end interviews with the mothers of malformed children. He did not specifically ask about drugs. But during one of the interviews that November, a woman in Hamburg told Lenz that she had taken thalidomide during her pregnancy, and had experienced peripheral neuritis. She had been very concerned about the baby. It was at that moment that Lenz, well aware of Contergan's success, first wondered if the drug was involved. He called several other physicians, expressed his concerns, and asked them to also make inquiries.

The next day Lenz reinterviewed several women, asking specifically about drug use, and four more women remembered having taken thalidomide during their pregnancy. One of them said that she hadn't originally mentioned Contergan because it was so insignificant. When Lenz inquired about thalidomide use by another woman who had delivered a child with phocomelia, the physician stated emphatically that he had not prescribed thalidomide but had prescribed Doriden instead. However, when Lenz examined the prescription slip, he found a note: "Doriden not available, replaced with Contergan."

Some 200 miles west of Grünenthal's headquarters, Anneliese Warren, the German wife of a Canadian military officer serving at a military base in Soest, saw one of the ads placed by Lenz and Schulte-Hillen that sought deformed infants. Since Schulte-Hillen had a car and Lenz didn't, the lawyer drove the doctor to the small town. A tall, slim, elegant woman with strong features and soft green eyes, Warren displayed little emotion as she welcomed Lenz and Schulte-Hillen into her home, but her immense courage was self-evident. ("She was so young and innocent," Lenz later recalled.) Early in her pregnancy she had been given medicine for nausea by a doctor off-base, she remembered clearly. Her baby had required immediate surgery when he was born because of stomach malformations. He was born with arms that were unnaturally short, four fingers on each hand, and feet that seemed attached to his hips. This was the first case of lower-body phocomelia that Lenz had heard of, and his sense of foreboding grew much stronger at the sight of this badly damaged little boy, whose name was Randy Warren.

By mid-November, Lenz had learned of fourteen cases of birth defects in which the mother had taken thalidomide. He called Dr. Mückter at Grünenthal and expressed his suspicions and concern. Lenz told Mückter that Contergan should be withdrawn immediately, and that every day the drug remained on the market was a deliberate experiment in human teratology. Lenz thought Mückter seemed nonchalant about the information, so the next day he followed up his telephone conversation with a letter to Grünenthal expressing his concerns about thalidomide. "Every month's delay in clarification," estimated Lenz in his letter, "means that fifty to one hundred horribly mutilated children will be born." The company ignored him.

November 16, 1961, was a historic day in the early story of thalidomide. While Widukind Lenz was writing to Grünenthal, halfway around the world William McBride met with H. O. Woodhouse, assistant sales manager at Distillers, and expressed his concerns about Distaval. He also wrote (again?) to the *Lancet*, and this time there was no doubt about whether the letter had been sent or received: it was printed in the journal the following month.

On Saturday, November 18, Lenz attended a pediatricians' association meeting, where Drs. Kosenow and Pfeiffer presented a paper on the causes of birth defects. After their presentation, Lenz stated publicly for the first time that "a certain substance" recently introduced in the West German market was responsible for the current epidemic.

He did not name thalidomide publicly, but after the meeting a pediatrician approached the professor. "Will you tell me confidentially," he asked, "is the drug Contergan? I ask because we have such a child ourselves, and my wife took Contergan." Lenz told his colleague that Contergan was the suspect drug, and before the meeting ended most of the physicians in attendance knew that thalidomide was the drug Lenz had referred to.

Three Grünenthal representatives, including the company's legal adviser, came to see Lenz the following day. They then all met together with the Hamburg health authorities, where Lenz presented his information. The Grünenthal people threatened to take legal action against the doctor for an unjustified attack on their company, but the government health authorities were alarmed, and asked Grünenthal to voluntarily withdraw the drug from the market. They refused. The same day, the company sent out almost 70,000 promotional leaflets to doctors across Germany declaring: "Contergan is a safe drug."

Before the end of that week Lenz again met with Grünenthal representatives, including their lawyer and a statistician, at the Ministry of the Interior; Lenz brought with him Karl Schulte-Hillen, the young lawyer with a deformed child of his own who had first contacted Lenz about the possible epidemic. But the Grünenthal people refused to continue the meeting in the presence of Schulte-Hillen, even though their own lawyer was present. Reluctantly, Schulte-Hillen withdrew from the meeting. Then the Grünenthal people attacked Lenz's reported fourteen cases, insisting that his evaluation was not statistically sound. The ministry officials were not impressed with Grünenthal's statistical attack, but they were very impressed with Lenz's report. They formally ordered Grünenthal to withdraw thalidomide from the market, or they would ban it.

The next day Grünenthal's executives met at Stolberg. They had before them Lenz's descriptions of fourteen cases of extremely rare birth defects coinciding with the use of thalidomide, and McBride's letter to Distillers. Together, these documents represented a terrible, ongoing threat to millions of families around the world. By this time, some of the executives realized there was no alternative but to withdraw thalidomide from the market. Keller, the inexperienced pharmacologist, acknowledged that "a teratogenic effect simply never occurred to me. It never occurred to anyone at Grünenthal." In personal agony, Keller later admitted, "I felt like a bus-driver who has run into a group of children and has killed and injured many of them." No similar remorse was ever expressed by other company executives.

Mückter dug in his heels. Despite the evidence that his drug caused horrific malformations, he refused to take it off the market. It is impossible to resist wondering why he would not agree to a complete withdrawal of the drug, given all that he knew by then. Did he truly believe these birth defects were isolated anomalies, statistical flukes? Or was he, like the army he had served seventeen years earlier, soldiering on in the face of certain defeat, with a similar contempt for the value of human life, out of loyalty to shareholders?

That seems like a stretch. But one has to note that, throughout the preceding six months, an even greater moral conflict between individual responsibility and blind loyalty was the leading topic of conversation: the trial of Adolf Eichmann had been broadcast worldwide on television since April, with a rapt and deeply divided audience in Germany. Eichmann testified personally about his central role in the Holocaust, at times boastfully taking credit for the Final Solution and the deaths of millions; nonetheless, to every criminal charge, he pleaded not guilty.

The trial in Jerusalem revived debates in Germany that had lain dormant since the Nuremberg trials fifteen years earlier. In fact,

Eichmann, the architect of Nazi genocide, retained the same defense attorney who had represented Göring, Ribbentrop, and Hess. More significantly, Eichmann resorted to the same defense: he had only obeyed orders. The implication—that individuals have no moral responsibility beyond obedience to their chain of command—was at the core of the banality of Eichmann's evil, in Hannah Arendt's enduring phrase. Ultimately, Mückter and the other manufacturers of thalidomide resorted to a similarly callous courtroom defense. In any case, throughout the time that Eichmann stubbornly insisted upon his innocence in the slaughter of millions, Mückter was fighting "to the bitter end" to vindicate thalidomide and to defend the rights of the company.

Whether or not the co-creator of thalidomide felt he was emulating the leaders he had served, the Eichmann trial was so much in the news that Mückter must at least have been haunted by the trial he himself might face. Most likely he was in a state of dread and denial, uniquely aware of how shoddy the company's testing had been and deeply conscious of the potential enormity of the disaster; his income, after all, was a percentage of thalidomide's swelling profits: about 30,000 DM ($8,000) a month, by one accounting—if sales were good. Whatever his reason, on November 25, 1961, Mückter refused to withdraw the drug from the market; he would only agree to a company letter warning of possible dangers.

But the next day a major newspaper, *Welt am Sonntag*, broke the news of Lenz's report, citing the most important passages of his letter: "Every month's delay in clarification means that fifty to one hundred horribly mutilated children will be born" because of thalidomide's unique reputation as a "safe" over-the-counter sedative. This was an enormous news story, whose terrible implications were instantly apparent to the whole country, and most especially to expectant mothers who had taken Contergan. Almost five years after the baby girl without ears was born on Christmas Day, her parents and the world finally began to learn why.

After reading the paper's headline and report, Mückter reluctantly agreed to withdraw thalidomide—but only from the German market. Grünenthal notified licensees in other countries of its decision, explaining it as merely a response to the "sensationalism of the

Welt am Sonntag story." Within days, newspapers, radio, and television were alerting women across Germany not to take the drug.

The following month, at about the time that Eichmann was found guilty of all charges, Grünenthal chemists set out for the first time to determine whether or not thalidomide crosses the placenta. They soon had an answer. It does.

Widukind Lenz wrote a paper for *Deutche Medizinische Wochenschrift* not long after describing some 161 cases of malformation where thalidomide was reportedly involved, as well as a letter in the *Lancet* endorsing McBride's letter. He also published the first paper that established the limits of exposure: women who took even one tablet of thalidomide between the twentieth and thirty-sixth day after conception (the thirty-fourth and fiftieth day after the start of their last menstrual cycle) were at risk of delivering malformed infants. At that time their periods would appear to be between five and twenty-one days late; very few mothers could feel sure they were pregnant that soon, especially in the years before home pregnancy tests. And yet, beyond that time, thalidomide caused no deformities at all. So there was another ghastly irony about this epidemic: although thalidomide's advertising eventually targeted pregnant women, only a few children were born deformed as a result. Most of the birth defects came from mothers who were not certain they were pregnant at the time they took the drug. This added considerably to the confusion and pain that followed, because women who were sure that they hadn't taken a drug after the first hint of pregnancy concluded that the deformities were genetic, and therefore also threatened their families.

At this point Grünenthal launched a withering barrage of personal attacks on Lenz with a quotation from Goethe: "Idiots and clever people are both equally harmless. Those halfwits and half-educated people who always recognize only half-truths alone are dangerous." Elsewhere the company's executives slandered Lenz's family. "His father was a famous and popular geneticist in Nazi times, since he had 'proven' the validity of the master race concept on genetic grounds," they told their counterparts from Richardson-Merrell. With these opening shots, the company began a siege on Lenz's integrity that lasted ten years.

It bears noting that neither the German government, the courts, nor the police stopped the bodily mutilation that this corporation was now willfully causing, day after day, around the world. It was a news story in the popular press that forced Grünenthal to stop selling thalidomide, and professional and medical journals that made public the evidence of the epidemic's cause. (More problematic was the German press's disregard for the victims.) In England, however, over the next fifteen years the thalidomide disaster would severely threaten the role of the free press in civil society.

On December 2, both the *Lancet* and the *British Medical Journal* announced that thalidomide had been withdrawn from the British market, in a letter signed by D. J. Hayman, managing director of Distillers (Biochemical) Company. He stated that "reports have been received from two overseas sources possibly associating thalidomide (Distaval) with harmful effects on the fetus in early pregnancy. Although the evidence . . . is circumstantial, and there have been no reports arising in Great Britain . . . we feel that we have no alternative but to withdraw the drug from the market immediately pending further investigation. . . . You may rest assured that the medical profession will be kept fully informed of developments."

Two weeks later, the *Lancet* printed William McBride's letter, and he became a popular hero. A Sydney newspaper subsequently named Dr. McBride Australia's Man of the Year for 1962, and in 1969 Queen Elizabeth made him a commander of the British Empire. The celebrity elevated McBride to the status of a household name in Australia.

The six-month-old infant named Randy Warren had blond hair and a slightly quizzical expression. Lenz could see that his severely malformed hips were going to require surgery, just to keep his spine structurally sound. In fact, since he was a rambunctious, playful child, he was continuously displacing his hips while rolling on the floor; again and again he was put into body casts. Since the Warrens lived on a small military salary, their medical options were limited. The most promising surgical techniques they could find were in Montréal, at the Shriners' Hospital for Crippled Children.

The Shriners, in all their years of philanthropy, had never helped a baby before, but they helped this one, after his parents delivered him there and returned to the Canadian military base in Germany.

Shriners' Hospital became Randy's home, off and on, for the next fifteen years. Ultimately he underwent thirty-two surgeries, many of which were geared to fitting him with artificial limbs. At one point, when his parents visited him, he didn't know who they were.

West German physicians estimated that about 40,000 people suffered from peripheral neuritis induced by the drug. Grünenthal's estimate was 4,000 cases. Ultimately, 8,000 to 12,000 infants were deformed by thalidomide, of whom 5,000 survived past childhood.

THE UNITED STATES IN PERIL

3

If a little knowledge is a dangerous thing, then where is the man who has so much as to be out of danger?

—T. H. HUXLEY, *ELEMENTARY INSTRUCTION IN PHYSIOLOGY* (1877)

WILLIAM S. MERRELL AND COMPANY was a subsidiary of Vick Chemical Co., best known for Vicks VapoRub; in 1960 Vick changed its name to Richardson-Merrell (the name used here). The company had set March 1961 as the target date for releasing Kevadon—their trade name for thalidomide—in the United States. They had *10 million tablets* waiting to hit the biggest drug market in the world, and their labs were ramping up to manufacture millions more.

Richardson-Merrell expected that Kevadon, along with their new anticholesterol drug, Mer/29 (or Triparanol), would be the key to joining "the truly big time." It was later established that Mer/29 caused cataracts and other severe side effects in monkeys—and that Richardson-Merrell had been aware of this and had extensively altered their clinical data to conceal the fact. Eventually, two Richardson-Merrell scientists and one executive pleaded no contest to charges of deliberately making fraudulent statements to the FDA, and in the end the company paid out an estimated $200 million to some 500 civil litigants who could prove they had suffered irreparable damage from Triparanol, as well as an $80,000 fine. It seems Grünenthal had found the right partner.

How could Richardson-Merrell's executives feel so sure that Kevadon would receive FDA approval? Because existing laws held that, after an application had been submitted, the agency had sixty days in which to decide that the drug was safe for the proposed use; if the FDA did not respond, the drug was automatically approved. And there was no requirement whatsoever to demonstrate the drug's efficacy. Essentially, it was up to the FDA medical officer to prohibit the chemical on the basis of hard data, not a doubt or a hunch. In fact, there had been considerable fraternization in recent years between the agency and the pharmaceutical companies. According to one medical officer, on Tuesday nights at Georgetown's Rive Gauche restaurant, there was always a pharmaceutical company representative there to pick up the tab for any FDA official.

Still, the burden of proving the safety of a drug technically devolved on the pharmaceutical company, as it had since 1938, when the worst drug disaster in U.S. history to date was produced by "elixir of sulfanilamide." Invented by Harold Watkins, a chemist for Samuel E. Massengill Co., the compound contained the poisonous solvent diethylene glycol—what we know as antifreeze. It killed 107 people (including Watkins, who committed suicide). Thereafter the federal Food, Drug, and Cosmetic Act (the FDC) of 1938 required pharmaceutical manufacturers to provide evidence of their product's safety when they sought approval; it also gave the FDA the power to remove a drug from the market if it were shown to be unsafe.

But that was as far as the law went at the time. And, as phrased explicitly in the drug bill of 1958: "The law does not require that specialized drugs . . . shall be tested for their therapeutic effectiveness." So it remained possible under that law for a consumer to spend years using a medically useless treatment, thereby being denied effective care. As several legislators later noted, because a drug did not have to be proven effective, a drug company could legally market water as a drug. To examine the law as it stood, Senator Estes Kefauver (D-Tenn.), was preparing to hold Senate subcommittee hearings on the pricing and efficacy of drugs.

Richardson-Merrell's confidence in their new product was understandable, given the apparent track record of the drug. As indicated in their application, after three years on the market in Europe and around the world, thalidomide had by this time been taken by tens of thousands of people on the basis of its advertised safety. And Grünenthal had never acknowledged any of the 1,600 reports of nerve damage and other side effects they had received from concerned doctors by the end of 1960. In fact, what they were doing amounted, arguably, to the largest uncontrolled drug trial in history, with no records whatsoever of who had ingested the millions of pills.

The application was submitted on September 8, 1960, and Richardson-Merrell executives confidently prepared to release the drug on March 6, 1961, with a massive promotional campaign. The campaign recommended thalidomide for anxiety associated with a fantastic variety of conditions that included abdominal pain, alcoholism, anorexia, asthma, cancer, cardiovascular disease, dental procedures, emotional instability, functional bowel distress, kidney disease, marital discord, menopause, nausea and vomiting, nervous exhaustion, nightmares, poor school work, premature ejaculation, and tuberculosis. As the *Sunday Times* put it dryly, "If thalidomide was not an elixir to cure all the ills of mankind, then it was certainly intended to make them incomparably easier to bear."

Unless the FDA's medical officer found fault with the application, thalidomide was poised to enter the enormous U.S. market. Further, planning was in progress for the drug to become, within two years, the first sedative ever sold in the United States without a prescription.

The pharmaceutical company had one last reason to feel cocky about FDA approval: they knew the ropes in Washington and they knew all the players, starting with the man at the top of the Food and Drug Administration. George P. Larrick had been commissioner for longer than anyone could remember, having started out as an FDA inspector in Ohio in 1923. After thirty-nine years with the agency, he was everything the drug companies could have hoped for.

By this time Larrick was in failing health, both physical and political. As one observer noted delicately, "It is no longer easy for Mr. Larrick to adjust to rapidly escalating standards of performance." The FDA had also just weathered an ugly kickback scandal involving Dr. Henry Welch, chief of the Antibiotics Division, and a close personal friend of Larrick's. Welch had collected more than a quarter of a million dollars in private fees from the antibiotics industry *while* he was certifying the efficacy and safety of their medications. Congress heard convincing testimony that similar relationships "were constantly invited, through repeated visiting and telephoning of FDA officers by drug company agents."

Richardson-Merrell, then, was operating in a decaying administrative culture governed by rituals of bribery that were both endemic and sophisticated, at a time when officials were not even required to make financial disclosures. The congressional spotlight on drug companies after the Welch scandal was irritating, but it would pass, and the old-boy network would surely prevail. In this instance, at least, Richardson-Merrell's executives paid more attention to the science of lobbying than to the science of medicine.

That was unfortunate, because if the company had paid attention to their own spotty testing, they would have realized that thalidomide could, in fact, be lethal when taken in syrup form: administered to eleven rats, the drug killed all six females; administered to thirty males, it killed twenty-three. When administered to a third group of rats, it killed all of them. But Richardson-Merrell performed those tests after submitting their application, and did not report them.

Clinical trials, under existing federal codes, required no FDA approval whatsoever. So, without any oversight, Richardson-Merrell had begun distributing thalidomide in the United States for clinical trials a year and a half before their application, and expanded the trials three months later to include pregnant women. This "investigational program" was not even organized by the company's medical department, as was usual, but instead by the sales and marketing division.

The company might have conducted cautious experiments with small samples of carefully monitored subjects. Instead, they distributed more than two and a half *million* tablets to approximately 20,000 patients, handed out by 1,267 physicians; the pills, of different colors and sizes, were distributed in envelopes or boxes marked only with the directions for use. This was dramatically larger than any previous drug trial conducted in the United States. The FDA later estimated that 3,760 women of childbearing age took the drug, of whom 207 were known to be pregnant. And it only took one tablet at the wrong time to cause horrendous birth defects.

There were hardly any FDA guidelines for collecting data during trials, so Richardson-Merrell sought none. The company's Hospital Clinical Program noted: "Bear in mind, these are not *basic* clinical research studies. We have firmly established the safety, dosage, and usefulness of Kevadon by both foreign and U.S. laboratory and clinical studies. This program is designed to gain widespread *confirmation* of its usefulness in a variety of hospitalized patients." Doctors "need not report results if they don't want to. . . . We may send them report forms or reminder letters, but these are strictly reminders and they need not reply."

Such was the extent of the human testing performed with thalidomide before its intended introduction en masse across the country. In fact, it could be said with a straight face that the most reliable test so far anywhere had been the "jiggle cage."

As the science editor for the *Saturday Review* wrote in 1962, FDA regulations left drug companies free to "begin using drugs on humans before safety has been established through animal tests, and they have the privilege of keeping the patients in ignorance throughout, lest knowledge of their guinea pig status have some undesirable psychological effect on the results of the experiment." Interestingly, this was a violation of the Nuremberg Medical Code, drafted after the trial of "The Twenty-Three" Nazi doctors. The informed consent of a patient in a clinical trial is now something we take for granted. But then, so too did the uninformed and nonconsenting patients of the 1960s, in all likelihood.

Patients do not expect they are joining a clinical trial when they take the pills their doctors give them.

Many of the participating physicians did not even record the names of patients who had taken thalidomide, much less their response to it, and the company even failed to make note of all the doctors to whom they had sent samples. Dr. Ray Nulsen, an obstetrician near Richardson-Merrell's headquarters in Cincinnati, Ohio, delivered two badly deformed children just weeks apart, one without arms or legs. All together, he delivered three living and two dead thalidomide children between January and May 1961. But Dr. Nulsen never reported this information; when it was later uncovered by investigators, Nulsen emphatically denied having given these patients thalidomide—and there were no records to prove different. Dr. Nulsen had also written numerous testimonials for drugs in the past, presumably to supplement his income. It turned out Nulsen had actually been a fraternity brother of one of the Merrell executives, and his testimony later showed that his laudatory articles were mostly drafted by Richardson-Merrell's director of medical research; the results of the company's "testing" were conveyed to Nulsen by phone, "or it may have been when we played golf." These early, highly favorable reports of Nulsen's about his "clinical trials" had been included in Richardson-Merrell's application.

The newly appointed medical officer assigned to review Richardson-Merrell's application was a country doctor, an academic, and, it was thought, a neophyte in Washington, unschooled in the realities of government health. She was also a woman. Dr. Frances O. Kelsey was forty-seven years old, married, with two daughters. Canadian-born, "Frankie," as she was called in the family, had studied zoology at McGill University in Montréal, and then earned a Ph.D. in pharmacology and an M.D. at the University of Chicago, where she also taught.

Kelsey's principal professor in Chicago was Dr. E. M. K. Geiling, one of the founders of pharmacology in America, who had himself worked under "the father of American pharmacology," Dr. John J. Abel. In the late 1930s, assisting Geiling, Kelsey contributed research for his study of the "elixir of Sulfanilamide" and its poisonous properties. This involvement with the worst drug-related disaster in America to date gave her a unique insight into the very real threat of pharmacological poisoning.

It was also in Chicago that she met and married Dr. F. Ellis Kelsey and became a naturalized U.S. citizen. During the war, while Heinrich Mückter was developing viruses with which to kill people, Frances Kelsey was studying ways to save soldiers' lives: she and her husband worked on the government's malaria project. Together they published a paper about their research with quinine and pregnant rabbits. What they had learned was that, while the mother's liver may break down quinine rapidly, the fetus's liver is unable to do the same thing, so the drug remains in the fetus's system much longer and is therefore toxic to the fetus but not the mother. She became, as she put it, "particularly conscious of the fact that the fetus or newborn may be, pharmacologically, an entirely different organism from the adult."

She and her husband, who was head of physiology and pharmacology at the University of South Dakota, collaborated on an important and enduring textbook, *Essentials of Pharmacology*, before settling for eight years in Vermillion, South Dakota. There, as a family doctor, Kelsey was fondly remembered as "a friend to over 50,000 South Dakota residents" in eight counties. A short, slim, unpretentious woman with a sweet smile and a slight overbite, her shoulders drooped forward a bit too much, and she was almost certainly not comfortable in a formal gown. But that was what she would have to wear from time to time after her husband was appointed special assistant to the surgeon general, as well as consultant pharmacologist with the FDA. The Kelseys moved to Washington, D.C.

Frances Kelsey found work as an editorial associate for the AMA. In that capacity she and her colleagues were accustomed to receiving and reading doctors' testimonials for different drugs, and some of those doctors came to be known as well-paid hacks for the pharmaceutical industry. If a paper by one of these doctors came across her desk, praising a new drug, Kelsey was immediately suspicious and generally did not consider it for publication in the AMA's *Journal*.

Her unassuming manner was distinctly British in style, along with her short, straight hair and flat heels. So, too, was her cooking, for which, she admitted candidly, she took pride in how few complaints of indigestion she received from her family. In an era when most wives' lives still revolved around the kitchen, Kelsey was out of the ordinary. But her well-mannered, business-like attitude did not suggest a troublemaker.

That may be exactly why the Larrick administration selected her for the same job formerly held by Dr. Barbara Moulton, who, together with the Welch scandal, had almost brought Larrick down. In the eyes of the agency, Moulton was a first-class troublemaker. She was an independent-minded woman who, it seemed, always questioned authority. She bore a slight resemblance to Eleanor Roosevelt (then serving with the United Nations) whom Moulton must have admired. She had repeatedly rejected applications for drugs whose safety she deemed unproven, only to be overruled by her immediate superior, Dr. Ralph Smith. Once, Smith had actually burst into her office while she was meeting with four drug company representatives, and, without waiting for pharmacological tests, he summarily approved their application. That may explain why four drugs—including Richardson-Merrell's Mer/29—had to be yanked off the market the same year, after numerous deaths and poisonings. Moulton was such an abrasive presence that Commissioner Larrick finally had her transferred to the Federal Trade Commission—to placate a pharmaceutical company, she was later told. With that revelation, Barbara Moulton resigned in disgust, and went directly to Senator Estes Kefauver.

Senator Kefauver, the long, lean Democrat from Tennessee who had been Adlai Stevenson's vice-presidential running mate in 1956, was chairman of the Senate Subcommittee on Antitrust and Monopoly. He was preparing to hold hearings to investigate price rigging in the drug industry, and the companies' incessant promotion of brand names to doctors who might otherwise prescribe generic drugs. (Forty years later, doctors are not the only targets of this advertising: now the pharmaceutical industry spends $2 billion a year promoting them directly to consumers.)

In the Kefauver drug hearings that followed, Moulton had the nation's attention as she blasted Larrick. "The Commissioner is a man with neither legal nor scientific training . . . [and] with no particular background in the field other than years of association with it. The *next* Commissioner should be chosen for intellectual integrity, courage, and professional attainments, so that he can command the respect and trust of his colleagues. In the present atmosphere, I seriously doubt whether any physician with a background in the pharmaceutical industry will meet these criteria." A sweeping indictment indeed. But, Moulton notwithstanding, Larrick could not be budged from his office. He had as firm a grip on the FDA as J. Edgar Hoover had on the FBI.

Kelsey met Moulton when she attended the Kefauver hearings to hear her predecessor's testimony. The two women soon became personal friends—a fact that cannot have escaped Richardson-Merrell's attention—and Frankie Kelsey's family welcomed Moulton into their home, while the rest of the FDA shunned her like the plague. At the same time, throughout the thalidomide ordeal, Moulton was a close adviser to Kelsey, along with her superior, Dr. Irwin Siegel, deputy director of the medical division, whom Kelsey remembers as her closest adviser. Kelsey, medical officer Dr. John Nestor (who was assigned Mer/29 for the agency), and Dr. Siegel were considered the "Young Turks" of the agency, who planned to raise the FDA's scientific standards. When asked if that was an accurate description, Dr. Kelsey demurred, observing that, at forty-eight, she could not be considered "young"; she did not, however, dispute the overall assessment.

As soon as she was appointed medical officer, Kelsey moved into her office. It was not, as one might expect, a paneled study in a stone building with high ceilings and thick carpets, but a "bare-floored room in a rickety barracks" on the Washington Mall (where the Museum of Science and Industry now stands); a prefab structure that had housed the FDA "temporarily" for fifteen years, wrote one visitor. "The walls are tattered with faded green paint; there are blinds but no curtains at the windows; desks and tables are piled with books and papers, in the manner of a place accustomed to long and intense concentration." But, as the *New York*

Times later remarked, Dr. Kelsey's story "points up the fact that all important government decisions do not flow from plush offices with wall-to-wall carpeting."

The application for thalidomide landed on her desk one week after she reported to work. In those days the agency had 7 full-time medical officers and 4 part-time (today there are 200). Although applications were usually assigned by rotation, she said recently, "Since I was new, they selected an easy one for me." This wasn't meant facetiously: the drug had been sold over the counter for years in parts of Europe, and its safety was thought to be unquestionable.

The first thing Kelsey noticed as she examined the four-volume application from Richardson-Merrell was the names of the doctors—including that of Dr. Ray Nulsen of Cincinnati, Ohio—whose testimonials were included with the application: many were the same hacks she remembered from her time at the AMA. According to established practice, certain doctors would write highly favorable reports of drugs without pay, and then earned good money on the reprints, which the drug companies paid for generously and then circulated by the tens of thousands. But the mere presence of those names did not bode well for the application in Frances Kelsey's eyes. In fact, as she looked through the application, she found it wanting in many respects.

The FDA pharmacologist who examined the application "felt that the chronic toxicity studies had not run for a sufficient length of time," Kelsey recalled recently. "He also felt that there were inadequate absorption and excretion data. The chemist found all sorts of problems or shortcomings with the manufacturing controls." It was disturbing that humans responded to thalidomide by lapsing into a "deep, natural sleep," but rats did not. The fact that no lethal dose for rats could be found seemed doubly disturbing to Kelsey, rather than reassuring—because it suggested to Kelsey that the rats simply weren't absorbing the medicine; since humans *were* absorbing the drug (as known from its sedative effect), the drug could still have toxic effects that couldn't be predicted by rat studies. So, among many other things, she demanded that the company prove to her that rats were indeed absorbing the drug: that the thalidomide was actually entering the bloodstream, and not passing straight through them.

It was also quickly clear to her that the animal studies and clinical trials submitted were insufficiently detailed and lacking in documentation, and, as a result, she wrote to Richardson-Merrell that "no evaluation can be made of the safety of the drug when used for a prolonged period of time."

Still, this did not provide Kelsey with grounds for rejecting the application. Furthermore, she was told by the drug company that no rats had ever died from thalidomide—whereas the public's use of dangerous barbiturates was growing by the month. She calmly informed Richardson-Merrell that she needed much better data and more of it before she could take any action. And she did something else unusual: she waited until the fifty-eighth day after the submission before declaring the application incomplete, and thus ineligible for submission—as if the application had never been submitted at all. That meant it would have to be resubmitted, giving her a further sixty days in which to ponder the application—and perhaps, to wait for any problems to be reported from Europe. It was a delaying tactic, but under the existing federal codes, it was about all that Kelsey could do. Besides, there was the possibility, at least, that at any moment Dr. Ralph Smith might burst into her office and summarily approve the drug, as he had done with Moulton. Larrick might even try to transfer her out of the FDA, but since the thalidomide application was only her second assignment, that might be a little too obvious for even the commissioner to risk.

In fact, Kelsey was troubled as much by the brazen inadequacy of the test data submitted as she was by its scientific sloppiness; Richardson-Merrell, apparently, had presumed that it would not be studied carefully. And the test results were peculiar, even intriguing. To begin with, why, wondered Kelsey, was it a potent hypnotic in humans, but did not induce sleep in animals? What did that suggest about its mechanism of action? That is, which of its biochemical properties made it work, and how?

And should a drug with unknown risks be marketed as a sedative throughout the United States on the basis of tests with a "jiggle cage"?

It was Frances Kelsey's husband, Ellis, who, as assistant to the surgeon general and an adjunct pharmacologist for the FDA, was entitled to write the agency's evaluation of Richardson-Merrell's submission, and he was unsparing. He described one section of their application as "an interesting collection of meaningless, pseudoscientific jargon, apparently intended to impress chemically unsophisticated readers. The selection of one chemical difference between two compounds as 'the most important difference' is absurd. What is the most important difference between an apple and an orange?" Because of Richardson-Merrell's deliberate disregard for elementary concepts of pharmacology, he went on, "I cannot believe this to be honest incompetence."

From the start it seemed to the executives at Richardson-Merrell that Dr. Kelsey was trying to hold up the application, and that her demands for more clinical data were unreasonable. But when Richardson-Merrell pressured her, the medical officer did not budge.

Richardson-Merrell subjected Kelsey to a withering siege of professional aggravation, provocation, and intimidation. Altogether she chronicled fifty-one exchanges with the company, when there should have been none: these contacts violated a specific pledge Commissioner Larrick had reluctantly made to Senator Kefauver during the recent hearings, that pharmaceutical companies would not be permitted to contact the examiners. How did Richardson-Merrell executives even learn the name of the medical officer assigned? It was Kelsey's superior, Dr. Smith, who gave Richardson-Merrell her name and telephone number.

Dr. F. Joseph Murray, the drug company's Director of Scientific Relations, began by acting cooperative with Kelsey, asking what data she needed, and what rewording she might require for the label's instructions to the patient. With some of the federally mandated changes she was insisting on, he tried to submit revisions casually, over the phone, with no paper trail. At other times he tried to sway her with data that he presented to her privately but that he omitted from the formal application. When Kelsey rejected the application as incomplete and "unsubmitted for the second time," Murray began trying to bully her. He pleaded to release the drug on March 6 as planned and to shift the burden to her, as a newly recruited medical officer. Richardson-Merrell was through with being polite with Kelsey. Perhaps they thought they could have more influence with her superiors.

Just ten days before the Richardson-Merrell target date for releasing thalidomide throughout the United States, Kelsey was looking through the latest copy of the *British Medical Journal*. "It was a little late in getting to us, because the mail was on strike," Kelsey recalls. "It was actually a shipping strike, I think, and we did not get our British and other foreign publications by air mail in those days. However, this did come in time." That was when she ran across the letter from Dr. A. Leslie Florence—the first case reports of peripheral neuropathy.

That letter changed everything. On February 23, 1961, Kelsey wrote to Murray at Richardson-Merrell that, since clinical reports of nerve damage had surfaced in Europe—reports that Richardson-Merrell had never acknowledged to her—she insisted that the company would have to disprove those reports. "We are much concerned," she added sharply, "that evidence of nerve damage was known to you, but not forthrightly disclosed in the application." Was it possible that the drug company hadn't seen the new label warnings added the previous month by Distillers or Grünenthal before they sent in their application?

By coincidence, Murray phoned later the same day—to let Kelsey know that the company would seek approval for over-the-counter sales. He did not expect her to have seen the *British Medical Journal*. Caught by surprise, Murray admitted that the company was aware of these reports and that they now planned to do exactly what Distillers was doing in England: to add warnings of possible nerve damage to thalidomide's package insert. In fact, Murray pressed her for a verbal commitment that such a warning would be enough to win approval. Kelsey dissented, noting that this was not an indispensable, life-saving drug, but a sedative, and therefore should not be marketed unless it were free of serious side effects. "I had the feeling throughout," Kelsey said later, "that they were at no time being wholly frank with me." She informed Richardson-Merrell that the application was still inadequate and that more detailed case reports would be necessary to address the peripheral neuritis issue. Since thalidomide was not a critical drug that could save lives, she wrote, "untoward reactions would be highly inexcusable." Forty years later, in an interview, she emphasized how different her criteria would have been, had thalidomide been submitted as a life-saving drug rather than a sedative.

In spite of the nerve damage issue, Richardson-Merrell, following the lead of Distillers in the British Commonwealth, was planning to declare on the U.S. label that thalidomide was safe to take during pregnancy; but they had not submitted to Kelsey a single iota of data to support such a claim. She warned Murray in writing that, given this proposed change, they would have to present much better information if they ever hoped to market the drug. In a letter to Richardson-Merrell, she enunciated the underlying principle of her job, which was only barely reflected in the existing law, but which would soon become the foundation of the reformed FDA. "The burden of proof that the drug is safe lies with the applicant," wrote Kelsey, "and must include adequate studies of all the manifestations of toxicity which medical or clinical experience suggest. In this connection we are very much concerned that apparently evidence with respect to the occurrence of peripheral neuritis in England was known to you but not forthrightly disclosed in the application." This she had taken as undeniable proof of the company's lack of candor.

By now Murray was under intense pressure from management, who thought he had "failed to get the application through by the usual methods and there should be some pressure exerted," as he told Dr. Smith; Murray went on to accuse Kelsey of libel. But the fact remained that, when Kelsey first raised the issue of nerve damage, Murray admitted that he had already read the reports. Smith did his best to placate the Richardson-Merrell executive, promising that the FDA would "reconsider" the demands that its Medical Officer was making for further documentation. Unassuaged,

Murray went over his head to Larrick's right-hand man to complain of the huge delays.

But Frances Kelsey stood her ground. In fact, she warned Richardson-Merrell she now had a whole new concern: since thalidomide could cause nerve damage in an adult, what evidence did Richardson-Merrell have to demonstrate that it would not more easily damage a fetus? Even though thalidomide was being sold in forty-six countries, it appears that Kelsey was the first person to pose this obvious question. But the answer lay in the fundamental mystery of thalidomide: by what chemical mechanism does it act upon the embryo to produce such specific malformations of the limbs?

Richardson-Merrell knew that it would take years of clinical trials to collect the data that Kelsey was now asking for, and the company was determined to release the drug by mid-November, their new target date, which had now been delayed for more than a year. They especially wanted to get those ten million pills out on the market by Christmas—the boom season for sedatives. They promptly offered assurances that the drug would not harm fetuses, but without any pertinent evidence. At the same time they bombarded Kelsey with documents about the dangers of *barbiturates* to the nervous system, and the threat they posed to innocent children and determined suicides. Thalidomide, Richardson-Merrell still insisted, would save lives.

As the battle dragged on, Kelsey forced Richardson-Merrell to re-submit their application a half-dozen times, while outside pressures upon her continued to mount. For her benefit, Richardson-Merrell staged meetings of "medical investigators" who insisted that thalidomide was safe—but they offered no new evidence, and not one of them had had any experience with thalidomide in pregnancy. In October, thirteen months after the original application, on the company's behalf Murray tried to coax Kelsey to set a deadline for approval; she would not. He asked if he could bring in the new labeling for her to consider informally; she refused.

Then, on November 18th, in Germany, Lenz made his first statement about birth defects caused by a popular new drug. And on November 29th, Richardson-Merrell was notified by Grünenthal that the drug had been removed from the German market.

"The company immediately phoned us, and told us the news, and said they didn't really believe it," Kelsey remembers, "but as a precaution they would put a halt to clinical studies going on in this country. But they did want to continue some that they had just started on its possible usefulness in cancer. That seemed no great problem to us, the benefit/risk ratio being entirely different."

Richardson-Merrell withdrew its application from the FDA on March 8, 1962, and removed Kevadon from the Canadian market—but only after the government in Ottawa demanded it. The company also informed Kelsey that they were notifying physicians connected with Richardson-Merrell in the U.S., warning them of the possible dangers from thalidomide. "There were some weird differences in wording of their two communications," Kelsey says, "that led us to think it might have been more widely used in this country than we had gathered from the new drug application, so we asked for a complete list of the doctors they had sent the drug to, and were very surprised at the numbers involved." Many of these physicians later claimed never to have received the letters, and most had not kept track of which subjects had taken the drug, so they could not even warn the patients of the potential danger.

In the absence of doctors' records, it can never be known how many babies died in the U.S. because of thalidomide's "clinical trials"; Dr. Lenz estimated that in forty percent of cases where there was fetal exposure, the infant died in its first year. Eleven women (or perhaps more) gave birth to thalidomide babies in the U.S., but there may have been many more whose parents never discovered that their children's malformations were caused by Kevadon. It is unbearable to speculate upon how many more might have been born but for the singular obduracy of Frances Kelsey.

In July 1962, *The Washington Post* published a front-page article by Morton Mintz, under the headline: HEROINE OF FDA KEEPS BAD DRUG OFF MARKET. It began, "This is the story of how the skepticism and stubbornness of a Government physician prevented what could have been an appalling American tragedy, the birth of hundreds or indeed thousands of armless and legless children."

On August 8, President Kennedy presented Dr. Kelsey with the President's Award for Distinguished Federal Civilian Service, for "standing off the promoters of thalidomide," as Senator Kefauver put it. "Dr. Kelsey's contribution," wrote the senator, "flows from a rare combination of factors: a knowledge of medicine, a knowledge of pharmacology, a keen intellect and inquiring mind, the imagination to connect apparently isolated bits of information, and the strength of character to resist strong pressures."

Kelsey invited Dr. Barbara Moulton to attend the ceremony: it was Moulton, after all, who had brought her concerns about FDA procedures to Kefauver even before thalidomide. And of course, Kelsey's whole family was present to meet the young President. "I felt like I was living in a fairy tale," she admitted later. Dr. Ralph Smith was there as well, on behalf of Commissioner Larrick—to salvage for the FDA whatever reflected glory he could from Frances Kelsey's moment in the sun. She became an American hero, gracing the cover of *Life* magazine, receiving and answering hundreds of letters, and then quietly returning to her job of protecting the public's health.

Two months before Richardson-Merrell withdrew its application, the renowned Professor of Pediatrics at Johns Hopkins University, Dr. Helen Taussig, whose pioneering heart surgery for "blue babies" in 1945 had transformed infant coronary care, welcomed to her clinic a young medical resident who was visiting from Germany. That was how she first learned of an "epidemic of infant monsterism." And a few weeks later Dr. Taussig herself flew to Germany and toured hospitals, visiting, studying, and photographing the thalidomide babies who had survived. Echoing the poet Heine, she wrote, "The one-third who are so deformed that they die may be the luckier ones."

Upon returning to Maryland, Taussig got in touch with one of her old pupils, who just happened to be Dr. John Nestor—Frances Kelsey's colleague and fellow "young Turk." One evening Dr. Taussig brought the photographs of deformed German babies to Kelsey and Nestor, and together they shared the same, dreadful thought: what if thalidomide *had* been approved fourteen months earlier? In

Taussig's opinion, "The marketing techniques of the pharmaceutical industry, which can saturate the country with a new drug almost as soon as it leaves the laboratory, would have enabled thalidomide to produce thousands of deformed infants in the U.S."

But whatever sense of relief there was quickly evaporated when Merrell admitted that their clinical tests had involved at least 20,000 patients—the greatest shock of all to Dr. Kelsey, she emphasized recently in an interview. Of those patients, the FDA determined, there were 3,760 women of child-bearing age, and 207 who knew they were pregnant at the time they took thalidomide. What was unknown at the time—and would remain so for almost forty years—was whether or not thalidomide could also be present in semen, and thereby produce the same destructive effect upon a fetus, by way of a sexually active male.

In the following months, Dr. Taussig played a critical role in alerting the U.S. medical community to the dangers of the two million tablets that Merrell had distributed in their "clinical trial." She helped track down doctors who had handed out thalidomide to the unsuspecting, and, where possible, located their patients as well. Where the patient was a woman who became pregnant, Dr. Taussig suggested X-rays to determine if the fetus was malformed. In such cases, she recommended therapeutic abortion, permitted in most states under the provisions that, until *Roe v. Wade*, were known as the "German measles law." "I wonder," she later wrote, "how long it would have taken to determine the cause of the malformations if the drug had produced some more common but less spectacular congenital defect."

On July 23, 1962, an FDA inspector visited Richardson-Merrell Co., trying to find how much thalidomide still remained unaccounted for in the U.S. Altogether, two tons of the material were never found. The inspector later commented, "I cannot help but have doubts about the adequacy and effectiveness of the [recall] procedures followed, since no formal letter was used (by the drug company), no material was returned, and Richardson-Merrell has no record or information as to how much material was destroyed or who destroyed it, if any." The missing thalidomide might have been dumped into the Ohio River, for all anyone knew.

In August, during a press conference, President Kennedy was asked what the government was doing to call in the supplies of thalidomide still unaccounted for. "The Food and Drug Administration have had nearly two hundred people working on this," Kennedy responded, "and every doctor, every hospital, every nurse has been notified. Every woman in this country, I think, must be aware that it is most important that they check their medicine cabinet and that they do not take this drug and that they turn it in." He then urged Congress to tighten existing drug laws, stressing that the thalidomide incident "emphasizes again the urgency of providing additional protection . . . to prevent even more disasters."

The FDA made a thorough investigation of Richardson-Merrell's "clinical trials," and eventually recommended that the Justice Department prosecute the pharmaceutical company for its unprofessional practices. But it ended there: the laws governing tests were too vague to prosecute Richardson-Merrell, and no case was brought against them.

By this time the country was engrossed in the drama of Sherri Finkbine of Phoenix, Arizona, mother of four and star of the children's show, *Romper Room*. While awaiting her next baby, she learned that the pills her husband, a high school teacher, had brought home to her from England were Distaval. Her doctor, along with other counselors, recommended abortion, and the Finkbines agreed. She sought a legal abortion, which, under Arizona law, was only permitted to save the life of the mother. Although her attorneys and doctor insisted that the situation was life-threatening to her, state and local officials warned that both the doctor and the patient would be prosecuted if the procedure was performed, and both would face two to five years in prison.

The Finkbines, whose lives were already devastated, went to the state's Superior Court, thereby allowing their story to become public; they were seeking a declaratory judgment that would bar any legal reprisals against them for having the abortion—which, at the last moment, was denied to her. In the end, Sherri Finkbine flew to Sweden, where she had the abortion. The obstetrician confirmed that the fetus was badly malformed.

"It would be the cruelest thing in the world to let my baby be born with only a 50–50 chance of being normal," Sherri Finkbine explained. "And I am concerned about our four other children. How would it affect them? Some people think that what I want to do is wrong. If it would make them happy, we would be glad to start again next month and try to have a normal baby." Thanks to her, a family in Pennsylvania, the Diamonds, finally understood how their son David had been injured before birth, and lodged the first civil case in the U.S. to go to trial. Nonetheless, Finkbine lost her job on television and began receiving anonymous death threats. In an account she wrote of her family's ordeal, entitled *The Lesser of Two Evils*, the archetypal mother of *Romper Room* firmly defended her decision, and her right to make it. (Her story was retold in the 1992 motion picture, *A Private Matter*, in which Finkbine's terrible dilemma was depicted by Sissy Spacek.)

A few months later, building upon grassroots sentiment raised by Sherri Finkbine's torment, Patricia Maginnis, Lana Phelan, and Rowenna Gerner started the Society for Humane Abortion in San Francisco and demanded, for the first time in the nation's history, the repeal of all abortion laws, openly challenging the law by providing education and information about contraception and abortion to women across the U.S. Then, in 1965, an epidemic of German measles (rubella) caused more than 15,000 birth defects in the U.S., and increased the pressure to establish the right of all women in the country to safe abortions.

The threat of drug-induced birth defects profoundly altered the whole context of the abortion debate and contributed significantly, a decade later, to the validation of all Americans' right to privacy by the U.S. Supreme Court, in *Roe v. Wade*. The thalidomide tragedy, and Mrs. Finkbine's in particular, had highlighted the fact that under existing state laws, even if she knew her unborn child had been severely damaged, an American woman had no right to chose whether or not to deliver the baby.

The true enormity of the disaster that was averted cannot be accurately estimated, but it is not hard to imagine the worst possible outcome. Given Richardson-Merrell's gigantic marketing plan for the first, "completely safe" over-the-counter sedative in the U.S.;

coupled with their absence of scientific discipline and, at first, their lack of corporate candor, which almost matched Grünenthal's; and given the absence of a national registry for birth deformities, and the utter inadequacy of the public health system for most Americans—Utopia notwithstanding—it is easy to picture a catastrophe that could have overwhelmed the country's resources for a decade or more.

That sort of speculation is certainly not to Frances Kelsey's taste. What she did was not, in her view, something to make a hoopla about. She was of course vastly relieved at the consequences that were averted, and grateful to President Kennedy for his consideration. And doubtless she didn't mind representing something positive to young women of the era who aspired to careers in medicine or in government. Although corruption had been shown to be endemic within the agency, that was not something about which she would ever share her opinion. The *Saturday Review* editor wrote, "She professes not to notice that anyone else in FDA is out of tune with her own sense of dedication to protection of the drug consumer."

As Dr. Taussig reminded the world in *Scientific American* that same August, "For most people, the story of thalidomide has ended. The tragedy will go on, however, for the infant victims of the 'harmless' sedative and their families for the rest of their lives."

The Aftermath | 4

Thalidomide makes people stare,
Thalidomide makes people care;
It stops me plaiting my own hair,
It even rules the clothes I wear.

—Catherine Purkis, age eleven,
thalidomide victim (1973)

The tragedy of thalidomide unfolded before a world aghast with horror. Along with the widespread insecurity of the Cold War era, it helped bring down the curtain upon the 1950s' vision of a science-borne Utopia. Between the threats of global nuclear annihilation and man-made carcinogens, it seemed as if mankind had released a scourge of new plagues, all of our own making. It seemed that way because that is exactly what had happened.

Throughout the 1960s and beyond, there was a growing recognition of the dangers of cigarette smoking, DDT, air pollution, and oil slicks. To this were added the disorders that a well-known journal sardonically refers to as "Diseases of Medical Progress"—maladies of our own making—of which thalidomide was the most infamous, but hardly the last. Eventually Agent Orange, acid rain, Three-Mile Island, Bhopal, and Chernobyl all contributed to the same lesson, learned the hard way: our species has a unique capacity for negligent destruction that extends far beyond the deliberate mayhem of our wars.

In the early 1960s the public was also learning about genetic mutations from the births of second-generation, postnuclear victims in Hiroshima and Nagasaki. Some of the birth defects suffered by thalidomide's victims seemed to resemble the malformations of mutation. In an era where Crick and Watson were celebrated for their discovery of DNA, and in which science-fiction films featured such genetically altered creatures as the giant ants in *Them* (1954), *Godzilla* (1956), and *The Fly* (1958), it was not surprising that people drew this connection: until thalidomide's mechanism of action became understood, explanations other than genetic mutation could only be guessed at. Was that what had happened to thalidomide's victims? Were they . . . mutants?

For many people, the simple, unspeakable word that came to mind at the sight of a disintegrated victim was "monster." In fact, the word teratogenic in ancient Greek meant "monster-making." The stunted, mangled limbs of the infants, whose photographs appeared in the press, suggested nothing so much as nature gone wrong, again and again. Many people recoiled from the sight of a thalidomide victim, insisting that this should not even be possible. It threatens our simplest sense of order, and vocabulary: is that . . . a hand? Is it polite to use the word "flipper"? (It is not, in ordinary conversation; but it is the word used in scientific discussion.) Is that still called a thumb if it has three joints? (It is.) This is a nightmare in flesh and blood. It is real, and it threatens us, by reflecting sunlight into the darkest corners of our fearful vulnerability. The sight of these babies' bodies, mangled as if by chemical shrapnel, proved unbearable to many, and not only strangers; nurses, doctors, and other health care providers as well as family members often turned away in revulsion from the victims, some of whom were happy babies, gurgling and gooing, wide-eyed, like most other infants.

These children posed puzzles about elementary medical practices that baffled even the most compassionate doctors. How, for example, do you monitor the blood pressure of an infant who has no arms or legs? What is a normal reading for that child? And when does a rising temperature threaten an infant who cannot disperse heat from the limbs?

"I have never seen such a reaction among my staff as when they were faced with thalidomide children," said Dr. Gerard Vaughan of Guy's Hospital, London. "They were horrified. I had great difficulty in getting them to carry out a psychological test or exam." And it was hard on many conscientious doctors to learn later that they had in good faith urged patients, like Randy Warren's mother, to take this poison early in pregnancy. These caregivers spent the rest of their lives in anguish for the part they had played in destroying dozens of lives.

The vignettes of horror that these infants and their families had to endure are themselves unbearable, like D-Day as recaptured in *Saving Private Ryan*; but the impact of this drug cannot be understood without some sense of the grief it produced. And these are only vignettes of childhood, not the hour-to-hour struggles these victims had to face for the rest of their lives. For many of these babies the future promised a world of hurt.

The greatest immediate anguish, of course, was reserved for the parents. Many were unable to bear the sight of their own children. When one realizes that some 5,000 thalidomide babies survived childhood—at least twice that many died from the drug—one must also grasp that each child represents a devastated *family*. In all likelihood thousands more were never identified as victims of the drug's malformations, because their parents and doctors did not connect the two. This was in part because of the range of deformities, and in some cases because of the wide variety of product names. The mothers may *never* have learned, even to this day, that their children's severe medical conditions were caused by a pill they had sampled once or twice. Before thalidomide, it had never occurred to ordinary people that a medicine could wreak such terrible havoc with a new life. So many families coped with gravely ill children for years without knowing it was because of deliberate time-saving, cost-cutting choices made by medical businessmen.

There was another reason that many parents never knew: some of the damage did not seem to fit any pattern. The majority of the victims' deformities involved foreshortened and distorted limbs, but others suffered from a wide variety of malformations, especially to the ears and eyes, as well as the genitals, which were sometimes missing, and an array of internal organs, particularly the intestines, kidneys, and heart. A common deformity was the complete absence of an anus.

The spectrum of the drug's destructive ability was much broader, and the suffering that resulted far greater, than that which was immediately obvious. One of the greatest continuing threats, without limbs, is the inability to disperse body heat. Most of the children were not expected to live beyond the age of ten, and many expired after a year or two or three. The best estimate is that some 5,000 victims survived childhood.

Many parents had the same reaction the first time they saw their child: they screamed. "They held him up for me," said one mother, "and he scared me half to death. He had no arms, except that on one side he had . . . something." The *Sunday Times* team recounted how, often, distressed nurses and hospital staff kept the babies hidden from their mothers, their bodies completely bundled up, until they left the hospital. So, in some cases the parents didn't have a clue until they got home, and undressed babies whose limbs were just an inch long, or whose ears were missing, like that first little girl born on Christmas Day.

Many fathers fainted away. At least one declared that if his wife brought "that monster" home, he would leave; when she brought the child home from the hospital, he abandoned them. Some fathers simply fled without a word, never to be seen again. So did some mothers. Others succumbed to years of intense clinical depression, and several to suicide. One woman had two deformed babies, born a year apart. For the most part, mothers took all the blame, from themselves as well as from others: blame for their prenatal behavior, their diets, and even the genes they had inherited.

In the event the mother had any relative who was handicapped, the genetic "responsibility" seemed self-evident.

Once the cause of the epidemic became public in November 1961, mothers could and did feel guilty about taking a medicine while they were pregnant, even though the vast majority did not know—and could not be sure—that they were pregnant when they took thalidomide (between the thirty-fourth and the fiftieth day after their last menstrual cycle).

How did parents endure this shock? The few who made it through without enormous collateral damage to their lives had to summon up the same enormous reserves of courage and devotion that are necessary to all parents of children with special needs and disabilities; then, perhaps, they needed still more courage, because of the special, peculiar horror that the sight of their children produced in even the most compassionate. Society does not reward such courage as those parents displayed, or even recognize it. Perhaps that is because those parents' experience represents our own worst nightmare, ever since we first imagined becoming parents ourselves. Of course, at the time of their children's birth, they had no idea what caused the malformations. At least, the first few thousand parents didn't. Toward the end, after thalidomide had been removed from the market, some mothers who had used the drug during the critical days were X-rayed, and then opted for abortion. But tens of thousands of couples endured months of horror instead of joy, knowing their infants might have been exposed, and awaiting their births with a fearsome dread instead of unbridled joy.

The impact upon the brothers and sisters of the newborn was no less horrific. This was the defining ordeal of their family life—leaving aside for now the crushing burden on their financial resources from that day forward. The operative emotions in most homes were guilt, shame, rage, and above all, terror—since the general presumption was that "it ran in the family," and the finest scientists in the field didn't have a clue. So most thalidomide families isolated themselves as completely as they could, and some simply did not admit to having the disabled kids they kept hidden. Harold Evans, editor of the *Sunday Times* of London, who visited hundreds of the children, wrote that their families had only "one common feature: they were in little camps on the outskirts of society." Many abandoned their deformed babies, formally or informally, banishing them forever to "homes" and spending the rest of their lives trying to forget them. "There was not one family I have seen that you could say was reasonably adjusted," said Dr. Vaughan. "Relationships were distorted by the experience, and few mothers escaped psychological trauma. The more subtle the damage—autism, epilepsy, mental retardation—the more alienated the families have become, destroying all efforts on their behalf. They are the greatest personal tragedies I have ever seen."

Others went further in their revulsion for the babies. "People said they shouldn't be allowed to live," recalled Dr. Claus Newman, a pediatrician in Roehampton, England. "But that was because they had a basic murderous attitude towards them. Excessive sympathy was mixed with fear and a threat. Faced with the sight of a disintegrated child, passers-by and professionals alike seemed to feel their own mental and physical wholeness at risk."

In Belgium, a couple named Van de Put murdered their severely deformed eight-day-old infant with poison. They later confessed, all the while insisting it had been in the best interest of the baby. And, to a courtroom in Liège, packed with 500 spectators, the doctor who signed the baby's death certificate admitted that he hadn't even had the courage to undress the infant. In a hushed, hoarse voice he declared, "If I had been the only one to know about the killing, I would have written, 'Death from natural causes.'" At that, the crowd applauded.

"If only my baby had also been mentally abnormal," said Suzanne Van de Put, "she would not have realized what her fate was. But she had a normal brain. She would have realized—she would have known."

When the couple was found not guilty of murder, the courtroom erupted in cheers.

In forty-six countries and a dozen languages, these tragic beginnings of life played out 10,000 times. And they continued to play out for another two years, because the pills were still available in some stores six months or more after *Welt am Sonntag* reported Lenz's findings to the world, three months after the *Lancet* and the *British Medical Journal* announced the teratological threat. In Brazil, because of confusion about all the brand names that included thalidomide, authorities seized two and a half million pills *six months* after the German withdrawal, along with a further 100 tons of the pure substance. It was mostly the same around the world. In Italy and in Japan, thalidomide remained on the market a further nine months.

Sweden was home to 105 damaged children, including 5 conceived after thalidomide was withdrawn by Grünenthal. Nonetheless, the government's medical board later ruled favorably, defending Astra, the manufacturer. The paternalistic medical board explained that they did not warn pregnant women about the threat immediately because "a warning could have caused increased psychic stress in those mothers who were already pregnant at the time in question, and might not have remembered the names of the various drugs taken earlier in their pregnancy."

The first civil court case brought by any parents of thalidomide victims began in December 1965, in the city of Södertälje. The parents' attorney, Henning Sjöström, had enormous difficulty finding expert witnesses who were willing to testify: they all came from the medical community, and none cared to cross Astra, the largest pharmaceutical company in Scandanavia. Experts in other countries had already been conscripted for the German criminal trial, where there were thousands more victims and so much more at stake; attorneys for the German families did not want their experts to offer their opinions in Sweden first. Finally, when most of the children were about five years old, a settlement was reached between Astra and the Swedish victims, who would share about $14 million over the coming years, more than they could have received under the law from a trial. No criminal verdict was rendered.

The fact that the United States and a few other countries had not approved thalidomide was extremely embarrassing to all the nations that had. It was an even greater embarrassment to the Free World that Communist East Germany, the German Democratic

Republic, had considered thalidomide and refused to allow it into their country: they found its scientific rationale wanting, and it "cannot be considered as an indispensable drug." In February 1962 Robert Kennedy visited Germany and denounced the Berlin Wall—which, on the Communist side, was being praised for keeping thalidomide out.

Randy Warren was nine years old when he learned that his injuries were caused by thalidomide: he read it on the medical chart at the foot of his bed. By that time the army had posted his father in London, Ontario, where, after one miscarriage, Anneliese had had two more children. The Warrens barely scraped by on army pay, but a settlement with Grünenthal was eagerly awaited—as it had been for eight years. Still, the family visited Randy as often as they could on Sundays from two to four, as allowed by the hospital. At Christmas he felt especially lucky, because he got a big bag of presents at the hospital and another big bag at home, the next time he was able to visit.

Otherwise, Randy's whole world was the hospital, and all of his friends were nurses. He hated men because the ones he met were usually doctors who operated on him, or wanted to. But he was close to some of the nurses, and sometimes visited their homes on weekends.

Thalidomide had only been on sale for ten months in Canada when Lenz made his revelation in November 1961; tragically, it took another four months and an outcry in the press before the government finally, reluctantly removed it from Canadian shelves. In April, five months after Grünenthal withdrew it, the head of the Food and Drug Directorate wrote to one doctor, "there is every possibility that thalidomide could indeed be reinstated."

The legal aftermath in Canada unfolded in two stages. As a preemptive move, Richardson-Merrell came to a quick settlement with ten families in Ontario, represented by Spangenberg Traci of Cleveland, which gave them $2 million to share; there is some question about how fairly that money was divided. But there were two important conditions: that the amount of money remain secret and, even more important, that Spangenberg Traci would not represent any more Canadian children, as they had avowed they were morally obligated to do. The strategy was simple and ugly: to keep other Canadian thalidomiders from even learning that they could sue until after the statute of limitations in their province expired. In Québec that was already a fait accompli: the statute expired twelve months *after the injury occurred*—when the twenty-six babies born there were about three months old.

But Arthur Raynes, the attorney who also represented the first U.S. victim, proposed a novel solution to those families: since Richardson-Merrell had a wholly owned subsidiary in New Jersey, he would sue them there, and prove that what had happened in Canada involved the interests of the state with one test case: an eleven-year-old from Québec, Denis Henry, whose cranial nerves had been damaged so he could neither laugh nor cry. He and his family sued Richardson-Merrell for "negligence and a conscious disregard for the safety of human beings." The pharmaceutical company replied that its legal obligation to shareholders would not allow for a settlement.

In the first battle, Raynes won the right to have the case heard in New Jersey. U.S. district court judge James A. Coolahan declared that "this court cannot realistically ignore the fact that the defendant's New Jersey activities played at least some part in the general thalidomide tragedy which occurred in Canada." As for the possibility that this test case would serve as a precedent for others, said Coolahan, "This court will not shatter that hope." But Raynes lost that victory in 1975, on an appeal by the company. When Raynes announced he would start a new suit in a New Jersey state court, Richardson-Merrell threw in the towel and agreed to settle. The twenty-six Québecois received $15 million U.S., with additional settlements for three other children. (Since Randy Warren was born in Germany, he did not qualify, of course.) Individual amounts ranged up to $999,999, because Richardson-Merrell would not pay any one plaintiff a million dollars, as a matter of "principle." Of course the law firm, Raynes, McCarty and Binder, kept their substantial portion of the settlement. Twenty-five years later, some of those victims in Québec are living hand to mouth, only marginally better off than their counterparts in Brazil.

The first civil case in the United States began in a federal court in Pennsylvania in February 1969, brought by Thomas Diamond and his wife, who had learned about thalidomide babies through the much-reported ordeal of Sherri Finkbine from *Romper Room*. During a trip Mrs. Diamond had been given thalidomide tablets by a hospital in Cleveland—ground zero for Richardson-Merrell's "clinical trials"—but she didn't have enough evidence to prove it. By the age of three, their son David's medical bills had already reached $20,000. This gives a good estimate of the pediatric medical costs for most surviving victims.

Richardson-Merrell, facing the first of an unknown number of U.S. lawsuits, and with its good name very much in the balance, was determined to win and thereby discourage other potential litigants. They argued that David was not a thalidomide child *and* that thalidomide was not dangerous. "There was a lot at stake for everyone," admitted the Diamonds' remarkable and committed attorney, Arthur Raynes, "and as a result it was a very bitter fight." For Raynes, it became a crusade against Richardson-Merrell that lasted more than a decade.

Since Mrs. Diamond could not prove that she had received thalidomide from the Cleveland clinic, she had a dim chance of proving the case, until a powerful witness arrived to speak on her behalf. Professor Widukind Lenz, who first announced the threat of the drug, came from Hamburg and testified that the damage David had suffered precisely matched that of other victims whose mothers had taken the drug in the same week of fetal development. As a result of the redoubtable Lenz's testimony and the growing publicity that the case was receiving, Richardson-Merrell settled out of court, suddenly and secretly, and decided to make their stand on the next case.

That was Shirley McCarrick of Los Angeles, a teenager who got pregnant in high school and delivered a severely deformed baby girl, Peggy, who had a short left leg that ended in a flipper, malformed hips, and severe internal problems. Shirley went to a high-powered law office that promised to look into the matter. Four years later not a single action had been taken, and soon after, the attorney in charge, who had been seriously ill, committed suicide.

Now just three months remained before the statute of limitations expired. Richardson-Merrell offered a settlement of $6,000. The young woman was financially desperate, and the company knew it. The new lawyer from that high-powered firm insisted she accept the offer; he knew she had no other way to pay him his legal fees. But, contemplating her daughter Peggy's future with dread, the young mother refused the offer. With that, the new lawyer emptied the box of her evidence onto her lap and walked out of the room.

In 1971, Shirley McCarrick's case went to court. The pharmaceutical company had hired every expert on teratology there was, it seemed, largely to prevent them from testifying for the victim. And even Lenz admitted under oath that her flipperlike leg was in some respects atypical. But Peggy herself was an irresistible, exuberant little girl. She appeared in court on her ninth birthday, wearing a pretty dress sewn specially by her grandmother, all in pink.

Richardson-Merrell was found negligent by the vote of the jury, ten to two, because the company had carelessly scattered millions of pills around the country for promotion, disguised as a "clinical trial." Peggy McCarrick and her mother were awarded $2.75 million—even more than her new lawyers had asked for. The judge limited that to $800,000 and ultimately, to avoid losing any further appeals, the company agreed to a settlement in secret, later reported to be in the vicinity of half a million dollars. But the outcome did not bode well for Richardson-Merrell, and helped persuade the company to settle the remaining nine cases in the United States.

In May 1968, after six years of legal investigation into Grünenthal's tactics in marketing thalidomide, the West German Ministry of Justice opened the criminal trial of nine executives of Chemie Grünenthal, including Mückter, Kunz, Keller, and Hermann Wirtz, one of the founders of the family business, who was now seventy-one years old. It was the largest trial in Europe since the 1947 Nuremberg trial of "The Twenty-Three" Nazi doctors, and when it began, it was an international media sensation, receiving almost as much attention as the Eichmann trial seven years before. The Grünenthal team could not have ignored the parallels, with victims, experts, and lawyers all condemning the decisions they had made; decisions, they insisted, they had had to make, because their only legal responsibility was to their shareholders.

The trial was held in the small town of Alsdorf, in a strange setting: a miner's casino and clubhouse that held 700 people. It was the largest available structure in the region, but it was still too small to accommodate the 400 plaintiffs with their families, the prosecution's 350 witnesses and 30 technical experts, and the hundreds of reporters, especially from other countries where thalidomide had been sold.

In all, there were 2,866 known German victims represented in the trial; a separate, more moderate group had settled out of court with the pharmaceutical company. After enormous planning and preparations to deal with traffic, lodgings, and the like, the trial began on May 27. The indictment alone was 972 pages long, based on 500,000 pages of documentation, including many seized in police raids on Grünenthal's "bunker" when executives did not comply voluntarily.

The co–prosecutor was none other than Karl Schulte-Hillen, both father and uncle to victims, the man whose visit to Lenz seven years earlier had triggered the initial search for the cause of the birth defects. Schulte-Hillen read the numerous charges, including the company's dishonesty regarding peripheral neuritis—sixty plaintiffs represented those victims. From the beginning it seemed as if some of the press were there on behalf of Grünenthal: names of witnesses, their addresses, and intimate facts about their lives appeared during the first week of the trial, presumably to deter others. It worked. Large numbers of scheduled witnesses never appeared in Alsdorf. But most of the families did, and three had no choice but to bring their children, the objects of the court case. So throughout the long trial, the young victims often spent whole days playing together outside the casino, while witnesses, jurors, and the accused walked in and out of the building.

The strategy of the defense appeared to be a war of attrition: to exhaust the opposition and the rest of the court, and, at the last moment, to make the best possible settlement, without ever admitting guilt. Their arguments and tactics were so heinous it is hard to believe they had the gall to put them forward. First, they insisted, through a long column of witnesses, that there was absolutely no evidence connecting their drug to either nerve damage or fetal malformations. After everything the world had experienced and all the changes produced by the invention of this compound, seven years later Chemie Grünenthal actually found scientists who insisted the drug was not to blame. Second, they declared that, even if such proof existed, the unborn baby—unless the victim of a criminal abortion—had no legal rights; ergo, it was not against the law for a company to cause such damage to a fetus, unless it could be proved that a criminal abortion was intended. Third, a variety of other factors might be responsible, including food additives, detergents, nuclear fallout, and even television rays. As well, they noted that no mechanism of action whatsoever had been offered by the prosecution to explain how thalidomide could cause birth defects.

Probably the most unforgettable argument put forward by the inventors of thalidomide was that all the babies' deformities were the fault of the mothers themselves, from botched attempts at abortions. And though it seems unbelievable, they offered an even more grotesque and far-fetched theory. One Professor Kloos from Berlin testified for the pharmaceutical corporation that thalidomide did not cause the fetus to be malformed; instead, he explained patiently, thalidomide had actually salvaged the lives of deformed fetuses that otherwise would have naturally miscarried. Grünenthal's expert was arguing that the "victims" were actually living abortions who had been *saved* by the drug.

Nothing the parents had experienced since the initial shock years before prepared them for this undeserved savagery. Of course, this trial would not in itself recompense the families one penny: that could only come from subsequent civil trials if Grünenthal were first found guilty of crimes. Many families sat in the courtroom each day, overcome, listening to these things being said about themselves and their damaged children, while some of the victims themselves were right outside, playing, reading, or lolling in wheelchairs, watched over by sisters from the Red Cross.

One of the witnesses who walked past the children was Dr. Heinrich Mückter, who took the stand, only briefly, at the start of the trial. "I would first like to say that I still regard the charge as a gross injustice to me personally," he said, fuming with moral outrage. After Mückter's testimony, an endless parade of experts began its way through the casino-turned-courtroom.

The first medical expert called by Schulte-Hillen was none other than Professor Widukind Lenz. Older, of course, but no less meticulous, Lenz had by now spent a large part of his professional career testifying on behalf of damaged children rather than caring for his patients in Hamburg. He felt a personal responsibility toward the victims, many of whom he continued to visit for years, and who came to regard him as a beloved family member. Of course, at the first scent of a lawsuit, many children had been brought forward with deformities for which thalidomide could not possibly have been responsible; it had often been Lenz's difficult and painful chore to make the distinction.

Lenz was cross-examined by no less than *eighteen* Grünenthal lawyers over twelve days without ever faltering or losing his calm, even-handed demeanor. "The only one who retains his self-control in Alsdorf," reported a local paper, "and refuses to be shaken out of his stoical calm by hostile questions, increasing insinuations about his knowledge, and the claim that his methods are 'useless' is Professor Lenz."

When, for example, the defense asked the pediatrician if it wasn't possible for the defects to have been caused by some unidentified virus, Lenz observed that "a virus would not stop at the Berlin Wall," a reply that baffled the defense attorneys, who hadn't even thought of that.

But fourteen months later, as this judicial juggernaut lurched forward, still far from conclusion, the defense won a surprising victory: they had Lenz's testimony expunged from the record as "an expert obsessed with his position, with an almost religious conviction of his mission," whose testimony had to be regarded as biased after he had "attacked his scientific opponents with arrogant lectures and insulting insinuations and criticisms. . . . Professor Lenz has no special value, but rather, judged by the requirements of objectivity, he is the most unsuitable expert imaginable."

Schulte-Hillen became almost apoplectic. "Professor Lenz has fought for years to prevent the truth from being suppressed by financial power." But Lenz himself was characteristically philosophical about the disgrace implied by the dismissal of his testimony. He acknowledged later that "my sympathy has not been equally shared between the company and the thalidomide victims, but I took great care not to be influenced thereby in my judgment of facts. . . . I decided to take it as a compliment to my moral engagement, rather than as an offense to my scientific honesty."

The pediatrician who had first pieced together the epidemic, who had spent years, mostly unpaid, visiting thalidomide-damaged infants like Randy Warren (who had made four difficult trips to Germany to be examined by experts for both sides in this trial), seeing more of them than anyone else—about 2,400 around the world, when he was done—and testifying on their behalf, was peremptorily dismissed. One of West Germany's best-known journalists, Walter Dirks, wrote in the *Frankfurter Hefte*: "We are currently witnessing a moral scandal. It has already been going on for some time. The scandal consists in the way the defense plays around with experts, the public prosecutor, the co-plaintiffs, and the court. . . . Doesn't the company realize that it has lost face by its behavior during this period? Doesn't it realize that the methods of its defense are bringing it into contempt?" The defense promptly brought this editorial to court and demanded a mistrial.

The elimination of Lenz's early testimony was a terrible blow to the prosecution, soon followed by another: their second star witness, Dr. Frances Kelsey, would not be allowed to testify in a foreign trial, forbidden to do so by FDA regulations.

The prosecution set about to prove that Contergan caused nerve damage and fetal malformation. Perhaps their best argument for the former was the fact that Grünenthal had admitted to paying out significant sums to neuropathy victims; but they also had testimony from many who hadn't recovered in seven years. As for the malformation, the proof was multifaceted and undeniable. To begin with, a graph of Contergan sales matched exactly—almost spookily—the recorded births of deformed babies, with exactly a nine-month lag. So when sales dropped dramatically owing to reports of nerve damage, so did the number of deformed infants. And there had not been a single case of phocomelia reported in Germany since nine months after the warning went out. Three hundred and fifty witnesses slowly hammered home the same message.

Meanwhile, Grünenthal's heavy-handed techniques continued, inside the courtroom and out. In May 1970, as the trial went into its third year, at least five journalists complained to the court that representatives of the company had threatened them with reprisals for what they had written. By this time, the defense had warned the victims' families that the trial (which they themselves were stretching out relentlessly) was delaying a civil settlement that would give the families the money they needed; and they went further. Grünenthal made an offer, tied to a threat: they would put up $27 million for the children; but if the criminal trial continued, they would fight to the end in civil court.

The offer reached $31 million, and the West German government added another $13 million—which represented the first meaningful assistance from the government. At that point, the organization of parents accepted on behalf of the 2,866 "Contergankinder."

But that did not automatically end the criminal proceeding. Day upon day the case crept forward. The trial had been in progress for more than two and a half years when Grünenthal argued that continuing the trial could bankrupt the company, leaving no money *at all* for the victims. In December 1970, two and a half years after they began, proceedings were suspended.

No guilty verdict was ever rendered, no personal responsibility was ever assigned, and no one went to prison. Since that was Grünenthal's goal all along, the company could boast they had won their war of attrition, and they did. They still do. The non–verdict in Germany is a source of lasting bitterness and grief to all the victims and their families. There was some consolation. The court did at least find that "a causal relationship has been proved" between thalidomide and nerve damage, and "there can be no doubt about the teratogenic properties of thalidomide in man, which were discovered by Professor Lenz. . . ." And the judges firmly emphasized that Mückter and his colleagues were neither being acquitted nor exonerated. The trial was simply being discontinued, "in the public interest."

The court published a balanced evaluation of the whole tragedy, endorsing the science that showed thalidomide was teratogenic, and stressing the need to change the whole system of development, promotion and sale of drugs, of legal control, and the attitude of doctors and patients. That was more important, the court found, than to punish a few scapegoats for the kinds of errors that society almost universally had permitted (or encouraged), and that might have occurred in any pharmaceutical company.

That is not what the evidence had shown.

5 Moral Justice and the Press

The Law is the true embodiment
Of everything that's excellent.
It has no kind of fault or flaw,
And I, my Lords, embody the Law.
—W. S. Gilbert, *Iolanthe* (1882)

The battle for the thalidomide victims of Great Britain lasted longer than the Trojan War.

It was a sprawling ordeal that had almost nothing in common with the cases in other countries. The issues it raised were fundamental assumptions about human society and British culture in particular, free enterprise and responsible capitalism, governance, politics, and most of all the law. But at its core the case was always about moral justice and the absence of it.

The structures of government in the United States are as fundamentally different from those of the United Kingdom as the rules of baseball are from those of cricket; not even the vocabulary is the same. This is no coincidence, of course, since U.S. law evolved out of and away from British law. And good riddance, some might say, were it not so easy to point out the gaping problems in U.S. jurisprudence. But in the instance of thalidomide, the U.S. system treated its eleven known victims—including David Diamond and Peggy McCarrick—as fairly as could have been hoped for, mainly as the result of powerful, expensive lawyering. And Richardson-Merrell was found guilty of negligence in a U.S. court for at least some fraction of their malfeasance.

England has no constitution, at least not in the American sense of a 5,000-word blueprint from which all other laws derive. The British constitution is the entire, overgrown thicket of law that "constitutes" the judicial system, including all legal precedents, right down to the required registration of dogs and cats. Since there is no separation of powers, instead of an independent Supreme Court there is a court of Parliament: a tribunal of the aristocracy within the House of Lords decides the fate of such cases that float to the top, unresolved in lower courts.

The Conservative Party governed England at the start of the thalidomide epidemic under the palsied hand of Harold Macmillan, an important postwar figure who had outlasted his own efficacy. The Tories' choice to lead the National Health Service is incongruous enough to guarantee a chuckle of disbelief, to those familiar with his name: Enoch Powell, the minister of health during the thalidomide crisis, was the fanatical Conservative who became England's most outspoken and unapologetic racist, railing against immigrants, Indians and Pakistanis in particular. It was somewhat as if President Eisenhower had appointed George Wallace as secretary of Health, Education, and Welfare. The idea of putting Powell in charge of Britain's health might have come from *Monty Python*, but this was five years before the comedy series began. Powell, with his dark, slicked-back hair, classic villain's mustache, and proud, vicious coldness, always came off like a martinet in search of an army. Unfortunately, he usually found one.

"The minister was quite sharp and said that anyone who took so much as an aspirin put himself at risk," recalled Christine Clark, one of a delegation of parents who met with Powell. "The minister expressed not one word of compassion or understanding." The minister of health was downright condescending, and apparently was only there to convey one message. "I hope you're not going to sue the government," Powell warned the parents. "No one can sue the government." He refused their request to warn the public about pills remaining in their medicine cabinets, as President Kennedy had done—he called that "foolish"; he would not meet with a thalidomide child; and in fact he refused to publicly acknowledge having met with the parents.

But Enoch Powell's most enduring failure as minister of health was his adamant refusal to launch a public inquiry. It probably was because he feared the victims' parents *would* sue the government; after all, many mothers were ordered to take the pills under the National Health Service, which paid for them, and Powell was close to the leaders of industry. In the same era much smaller tragedies received meticulous scrutiny by the government—the Aberfan coal tip disaster, for example, and every airplane or railway accident—but Parliament chose not to learn how an avoidable epidemic of birth defects had occurred across the country. In July 1962, when an investigation was proposed on thalidomide, the House of Commons voted it down.

It may seem reasonable to suppose that American readers could make sense of British jurisprudence if only the writers applied themselves, but in truth only a handful of British judges and barristers—the high priests of attorneydom—are credited with grasping all its vagaries, many of which came into operation in this case. White wigs are not the only oddities in a British courtroom. Given England's judicial eccentricities, the financial and political power of a giant industrial corporation, a complicitous government, and the transient lawyers of the hapless victims who repeatedly missed crucial opportunities, the picture that emerged of the parents' legal situation was bleaker than *Bleak House*.

The determinative fact was this: after the first lawsuit was filed against Distillers on November 7, 1962, the law forbade any mention of thalidomide whatsoever in the press; so there was no mention for the next ten years. The case was sealed in "a legal cocoon," as the *Sunday Times* Insight Team described it. And well they should know, because they alone continued to investigate the story throughout its duration; no other journalists in the world attempted to follow the proceedings or to tell the story for fifteen years.

The purpose of the gag order was to avoid "trial by newspaper": to assure that juries, perhaps years down the road, could not be influenced by the press rather than evidence. Newspapers almost always respected such gag orders rather than face charges of contempt of court that could result in imprisonment and fines. In this particular case the effect was to prevent many English victims from knowing there was a class-action suit, and so it was years before they all emerged.

Ironically, before the gag order, the popular press was doing a remarkable job of putting Distillers' best case forward for them. The *Times* of London, sister paper to the *Sunday Times*, published an article titled "Thalidomide Tests Showed No Sign of Danger," putting forth as factual the arguments that framed Distillers's legal defense. "The drug," the *Times* crowed, "was subjected in Britain with great thoroughness to all the tests which any pharmacologist would have applied in the circumstances. . . . " The article had been written by the *Times*' industrial reporter rather than its medical correspondent, for reasons that remain murky. It eventually came out that this and similar articles, as well as the government's position on the affair, had derived from an effective, high-level campaign by Distillers, one of the country's largest corporations. The reporter who wrote the article retired from the *Times* the day after it was printed, and went to work for Distillers.

This is only one of hundreds of examples of the power and wile employed by Distillers over the coming decade. The corporate behemoth already had overwhelming influence over the High Court, the House of Lords, and the Conservative government in the House of Commons—not to mention the charitable Society for the Aid of Thalidomide Children formed by Lady Hoare. Much of this staunch allegiance was unspoken; it didn't need to be explained. It derived from the philosophy shared by industry, the lords, and the Tories of that era: social Darwinism, unadorned and unapologetic, and any who fell by the wayside in society were the responsibility of charity. Those three institutions kept a fourth—the press—completely muzzled. Any one of them might have removed the gag, but none did. Her Majesty Queen Elizabeth II might have said something. She did not.

Eventually, 456 children in England were identified as victims (the numbers shifted continually over the years, as more victims came forward), thanks largely to Dr. R. W. Smithells of Liverpool, who had created a registry for birth defects in 1960 after seeing five children born without arms. He guessed this was somehow caused by a virus, but when Distaval was suddenly removed from the British market, his registry proved crucial in finding and informing victims' families. He also formed the Kevin Club for victims in northern England, named after one little boy. Smithells, like Lenz, remained an important figure in the children's lives, personally and scientifically, for many years. Sadly, though, he believed Distillers' propaganda, coming as it did from so many sources in the government, press, and industry, so he did not support litigation. Equally, the charitable trust established under the auspices of Lady Hoare, wife of London's lord mayor, was an arch-conservative organization that also firmly discouraged any lawsuits by its beneficiaries.

Another factor against the sixty-two test families who decided early on to sue Distillers was their reliance on public funding, controlled by the Law Society, to pay legal expenses. The society's outlook, as expressed by one of its presidents was unhelpful: "We have known virtually from the beginning that there was no case." The society determined—very slowly—which lawyers and experts could be paid for and which could not, based on the unrealistic standard of what a "person of reasonable means" would spend; whereas the families in this unreasonable situation were fighting a Goliath who would spare nothing to quash this nuisance and any noise it might emit.

It is pointless to recapitulate the first five years of the court case waged by those families because, tragically, they amounted to nothing. Under English civil law, the burden of proof lay with the families to establish that Distillers should have known that thalidomide could cause birth defects, and should have performed their own animal tests rather than rely upon Chemie Grünenthal's. Two professors at the University of Durham, Harvey Teff and Colin R. Munro, have noted that, "careless commission (or omission) will not in itself constitute negligence in law. . . . It has to be shown that the carelessness was an operative cause of the injury." The plaintiffs, then, had to prove that "properly conducted tests [for teratogenicity] would have revealed the dangers." That might have been done, even without discovering the chemical mechanism of action whereby the drug wreaked its havoc—a scientific undertaking that took another thirty-five years.

The families' solicitors and barristers in shifting succession failed their clients badly: they chose a poor test case and they barely studied the literature that would have challenged Distillers' claim that tests necessary to reveal the threat were not available. They failed to obtain experts; in England, many experts were not about to contradict powerful colleagues in the pharmaceutical industry. Most of the world's experts were committed to the criminal case in Germany and were not available for the British suit. So neither McBride nor Lenz were consulted. Perhaps as a result of their lack of foresight and preparation, the families' lawyers and their barrister, Desmond Ackner, made it clear to the families in 1967 that they could not win and should settle quickly for whatever they could get; it almost seemed as if Ackner had switched sides and was acting on Distillers' behalf against his own clients. Ackner presented the company's offer to the families by saying that anyone who did not accept "would deprive all the others of a settlement. . . . It was an all-or-nothing deal"; unless all the families agreed quickly, "a ghastly tragedy could result." As if nothing of the kind had happened yet.

"I was left feeling that we were the guilty parties for having taken the drug," said Clark.

But in retrospect, the *Sunday Times* Insight Team agreed with Ackner and his lawyers: had the case gone to court as scheduled in February 1968, "we believe that they would almost certainly have lost everything. They would have lost because their lawyers believed they could not win and had failed to mount the best case."

As it was, the sixty-two families capitulated to their barrister, and accepted 40 percent of the maximum they might have received from a trial. One hundred percent would have represented what the children's medical costs, special care, and lost income would be over a lifetime: that is, a wild guess. That was decided in the High Court in 1969. The families' actuary estimated that a child with no limbs at all would need $240,000 over a lifetime, a figure that gave no realistic allowance for the inflation that has in fact occurred since then. Under the settlement, such gravely injured children received a lump sum of $21,700, with lesser payments for less damaged children: sign here or else. The accounting was so complex that it was years before anyone objected that $21,700 did not amount to even 10 percent of $240,000. To appeal the settlement would automatically condemn to penury the other 340 families waiting for the case's resolution to sue Distillers themselves; by this time the statute of limitations was nearing expiration, leaving the families without hope of justice. As a gesture of compassion—and it was hardly that—Distillers established a small charitable trust fund of some $7.8 million, from which interest could be drawn annually and shared between the 430 or so victims then identified (some 20 more presented themselves in later years).

Despite the all-or-nothing clause, six families refused to settle; one did so out of conviction that their armless son had to learn to care for himself and to maintain his independence; another, out of similar pride, wouldn't allow a social worker to think her daughter lacked anything she needed. But the leader of these dissidents was the wealthy owner of an art gallery in Mayfair named David Mason, whose daughter Louise had been severely affected in all four limbs. He took up his cudgel for and with the other families, and insisted they would never accept such an insufficient offer.

At this point any vestige of sanity that was left to the proceedings vanished. The association of families tried to coerce the six dissident families by having them removed as legal guardians of their own children: to make them wards of the court. David Mason reacted with utter disbelief. "I was being taken to court by the solicitors originally engaged to act for me, to have my right to decide my daughter's future taken away from me. I thought, Can this really be British justice?"

Unbelievable but true, the court stripped the six families of legal guardianship of their own children.

This last, devastating judgment was perhaps the best indicator of the invincible conspiracy the families were up against. It was overruled on appeal a year later, in 1972, as the families' lives ticked away: by now the youngest of the children was ten years old, without a brass farthing from Distillers. Since all the families hadn't agreed to settle, the whole situation ground to a complete halt. There it sat, motionless and lifeless, for months. Apart from those involved in the litigation, no one in Britain knew a single thing about the case.

The *Sunday Times* first received a batch of 10,000 internal Distillers' documents in 1967 from Dr. Montagu Phillips, a pharmacologist on the defense team that had subpoenaed them. Phillips had copied them and now offered them to the newspaper, together with his expertise, for $20,000—roughly the amount the most damaged of the British children were eventually offered in settlement. Soon after, Swedish attorney Henning Sjöström brought the newspaper three suitcases filled with documents that Grünenthal had been forced to hand over to the German court for their upcoming trial—about which, incidentally, the *Sunday Times* would not be able to report either.

Now the paper's Insight Team moved into an office of its own and began studying and compiling all the documents. Again and again a variety of top-notch solicitors counseled the paper that they could not publish a word on the subject without risking imprisonment. Harold Evans, the paper's famously urbane editor, was committed to a variety of social causes, and the fate of the thalidomide children had definitely become one of them.

Evans had come to the *Sunday Times* in 1966 from the small Darlington paper, *The Northern Echo*, where he had published two stories about local families struck down by thalidomide, and had already made a large name for himself as a crusader. He was moved, first, by sympathy for the desperate, defeated people, then by his outrage at the inability to write about their hapless court case. Evans believed that one had to base a campaign upon the type of investigation that could only be conducted by a major newspaper. "If you're going to stand up to the bully boys of the world," he once remarked, "then you need the muscle of a big newspaper." And the simple principle that drove his determination was that "corporations have moral as well as legal responsibilities." So, soon after arriving at the *Sunday Times*, he launched an investigation. The owner of the paper, Lord Thompson, was prepared to support Evans—just so long as he did not attack the queen.

For years he tried to find a maneuver that would free the *Sunday Times* from the gag order without facing imprisonment himself, as other British editors had, notably during the Profumo scandal. Thus the thalidomide case was a secret scandal, and it was the job of the newspaper's attorney, in Evans's words, "to keep me out of jail long enough to make it a public one." Otherwise, he and his colleagues were, in his words, "silent witnesses to a moral outrage." And all along those who might have spoken out, first in Harold Wilson's Labour government, and now in Edward Heath's Conservative government, had remained completely silent on the subject.

At about this time, one of the actuaries who had helped determine the "quantum" of which the victims would receive 40 percent published two scholarly articles using the actual calculations in the case as examples; from these, the Insight Team repeated the calculations and realized that the children had actually received just half of the negligible amount the court's formula should have produced—*because of faulty arithmetic*.

After years of fact-finding and hesitation, Harold Evans could no longer restrain his burning sense of injustice, and on September 24, 1972, the *Sunday Times* published a detailed, three-page story by the team titled "Our Thalidomide Children: a National Shame," which included statistics, actuarial tables, and photographs of thalidomide victims. In their editorial, with the headline "Children of Our Conscience," the paper deplored the indecently small settlement of £3.25 million, when that year alone Distillers had had pretax profits of £64 million, with corporate assets of £421 million. The article made the larger point that civil courts could never assure a fair settlement between parties like this, with grossly unequal means. "The law," noted the editorial, "is not always the same as justice."

Importantly, what this article and its two immediate successors did not include were any facts derived from the Distillers' documents "borrowed" and copied by Phillips, which would have instantly given the pharmaceutical company grounds to stop publication. By skirting legal issues, this first article tested the attorney general's appetite for muzzling the press.

The *Sunday Times* reprinted their article and supplied copies to every member of Parliament and every major media outlet in the country. Jack Ashley, a very popular M.P. who had lost all his hearing in a botched operation, and who often championed the handicapped, phoned Harold Evans immediately and became a committed ally in Parliament. That was some comfort, since Evans had also just heard from the attorney general's office, warning that parts of the first article appeared to violate the gag order, and that contempt of court was a real possibility "in view of the reference already made to a future article on the subject."

Nonetheless, the following Sunday another article was printed, along with letters from several victims' parents. By now the other media in England were beginning to pick up on the story. Promising news came when the attorney general informed the *Sunday Times* that he would not take action for the first article. But at almost the same time, Distillers won an injunction against a third article that the paper had planned to publish, but would now hold off on printing until its appeal was heard by the High Court of Lord Chief Justice Lord Widgery and two other senior judges.

Before the High Court were the competing interests of a free press and the administration of law. The *Sunday Times* framed the entire debate in terms of the separate demands of moral justice and legal justice. This was the argument eighteenth-century philosopher David Hume had made, insisting that morality is a separate realm from the factual world: in Hume's simplest formulation, "ought" can never be derived from "is." Here was an argument with a pedigree.

During the appeal, Harold Evans stood tall in the witness stand, testifying to the "great public dismay that after ten years the law has not produced justice"; not even legal justice, with which the *Sunday Times* acknowledged it could not interfere. The paper was waging its campaign on behalf of moral justice, which had to begin with the recognition that, admirable as Lady Hoare's society was, this really was not a question of charity at all: the children were *owed* fair compensation. Evans could also show that even 100 percent of the "quantum" was utterly insufficient for the special needs of the victims throughout their difficult lives to come.

All three justices denied the appeal for the *Sunday Times*, arguing—to the extent they bothered—that such a precedent would allow less scrupulous publications, of which Britain has a few, to affect the outcomes of other cases. Thereby the highest court in the land banned the publication of the third article in the *Sunday Times*. The fact that the unpublished article made use, for the first time, of the documents obtained from Phillips was a separate issue, for which Distillers launched a separate lawsuit, to protect the confidentiality of its internal memos (Dr. Phillips had copied the thousands of documents that Distillers had been compelled by the court to give the families' lawyers, for whom Phillips had testified). Although the High Court forbid publication by the *Sunday Times* of facts about Distillers' testing, it failed to reprimand the drug company for making its argument public from the beginning, that all appropriate tests had been conducted on thalidomide.

Rejected by the High Court, Evans began the process of going to the Court of Appeal.

Jack Ashley, the deaf member of Parliament, was now gagged as well: he was chastised for repeatedly raising a matter in Commons that was *sub judice*. It was not by coincidence that his counterargument in the House of Commons mirrored the *Sunday Times* argument before the High Court: "Moral justice is not a question of law, and therefore cannot be *sub judice*." How could it be, he demanded, that Parliament was forbidden to discuss what the attorney general had allowed the *Sunday Times* to investigate? Ashley was incensed. "For 370 children, the sword of Damocles has been replaced by the jagged edge of a broken whisky bottle."

"The children's living standard is falling as the profits of the company are rising," Ashley went on. "We are witnessing not only a shabby spectacle but a grave national scandal, a display of national irresponsibility which has seldom or ever been surpassed." How extraordinary it must have been for the victims' families to finally hear words shouted in the House of Commons that for ten years they had only heard spoken in their own parlors—or in their heads, perhaps, late at night.

So a prolonged and superheated parliamentary debate was launched about the thalidomide children after Harold Wilson, now leading the opposition Labour Party, put it into the schedule. One member, a family doctor before being elected, recalled how Distillers had pressed him to prescribe Distaval. "There is an outcry in the country," he added, "and if the House of Commons is not a place we can shout about it when something is to be hidden behind a smokescreen of legal procedure, God help the ordinary people when it comes to claiming redress of their grievance." Prime Minister Heath continued to insist that this was simply an issue between a private business and its customers, in which government had no role to play. The only legal responsibility Distillers had, as far as the Britain's Conservative government could see, was to its 250,000 shareholders. At the same time, Sir Keith Joseph did announce that £3 million ($7.5 million) would be spent by the government to improve services for all handicapped children. It was a small victory for the families, but as David Mason put it, "Nobody is getting any glory out of this. At the end of the day, my daughter Louise still has no legs."

After being bottled up for so long, the whole painful story spilled across Britain, producing an outcry. Distillers sensed the public's growing outrage and tried to abate it by increasing their settlement to the children from £3.25 to £5 million ($12 million), but it was too late: Distillers was facing a public boycott. Now they could actually count, in pounds and pence, their diminishing sales of vodka, gin, and whisky.

One participant in the boycott remembered how the plight of the thalidomide children cut straight through political complexities to a simple question of compassion. Gilla Kaplan was a young graduate in biology, recently arrived from Jerusalem, and like many others who were involved, she had never participated in any public political activism. Still, years later, Kaplan would feel proud that she had taken a stand on this issue.

Across London posters began to show up mysteriously, overnight, on lampposts and Tube entrances; posters, for example, of English gentlemen in a pub with a whisky bottle marked "Thalidomide." There was no indication of who put up these posters; it appeared to be the work of a spontaneous collection of small, outraged groups. The attorney general went so far as to declare the posters illegal, and Scotland Yard set about tracking them down. This was proof, it was said, of what would happen if the *Sunday Times* were allowed to publish its article.

The boycott began to appear more like a well-orchestrated campaign, and it made the *Sunday Times* very uncomfortable, knowing full well that such tactics would enrage Distillers—which it did; but it also undermined Distillers' arguments for gagging the press. And it was intended to, by a top executive of the organization that actually had orchestrated the poster campaign in secret: *News of the World*, flagship of media tycoon Rupert Murdoch, getting onto the bandwagon very, very quietly. It seems the plan was to provide other newspapers with a photo opportunity where they could include the posters and get around the news gag that way. But no newspaper ever ran a photo of a poster—not even *News of the World*. The threat of contempt of court was too daunting.

Meanwhile, as the months went by, the parents' settlement and the situation with the six dissident families sat motionless, completely stalled.

What are the responsibilities of an individual who owns shares in a joint-stock company that causes a public disaster? Law school libraries have built annexes for all the books on that subject, but now the *Sunday Times* turned the nation's attention to that question, together with a handful of small investors in Distillers who felt compelled by conscience to take action.

In the cases already resolved in Germany, Sweden, and the United States, the directors of the different pharmaceutical companies repeatedly insisted that, in spite of their personal sympathy for the thalidomide children, their overriding responsibility was to their shareholders (though they also extended loyalty to their management in such trying circumstances). Distillers went further when it was challenged to contribute £20 million to the victims' fund. Its chairman, Sir Alexander McDonald, replied in writing. "Directors of a public company who acted in such a way and on such a scale might at once become subject to legal proceedings" from other shareholders. Although it was highly unlikely that Sir McDonald would go to jail if he helped the children his company's untested product had maimed, his reply raised an interesting and important question: just how did Distillers shareholders feel about the directors' handling of the thalidomide case?

In November 1972, ten Distillers' shareholders met, with only a few thousand shares between them of the 300 million shares outstanding (each worth about £2, or $4.80). They discussed their option, because they only had one: to call an extraordinary general meeting of shareholders. But to do that, they would need signatures from investors who owned 36 million shares. Since Britain requires that shareholders' names be public information, they began by purchasing the thirty-two-volume list of 250,000 shareholders: that alone cost almost $4,000, paid for by the *Sunday Times*. The majority of the shareholders listed owned 100 to 200 shares.

A reporter for the *Guardian* helped the families go through the volume marked "L-P," and found a giant life insurance company, Legal and General Assurance Company, which owned £6 million in Distillers' shares and was a large force in the financial district. After consideration, the chairman of Legal and General, Ron Peet, issued a press release: "The moral claim of the thalidomide parents makes a strong appeal to our sympathies and we, as shareholders, would support a more generous settlement. . . . " Peet later explained that "Personally I felt that Distillers were behaving irresponsibly. . . . The children had a strong moral claim. Here we had a public company involved in a disaster; they were not in a position to wash their hands of the matter. . . . Second, Distillers were behaving in such a manner that they were beginning to prejudice their own commercial interests. . . . Our own self-interest entitled us to make our views known."

With this statement from Legal and General, many townships that had funds invested in Distillers began to express solidarity and called for an emergency general meeting of shareholders; soon they were joined by thousands of independent insurance brokers, who spread the word among Distillers' investors. As this movement grew, Distillers suddenly upped its offer to about £11 million—but that included a scheme whereby the government would forgo taxing the company on those monies, which meant British taxpayers would pay the increase in Distillers' offer. Distillers claimed publicly that the government had given the company reason to believe it would accept such an arrangement, as had been done in Sweden and Germany. But the chancellor of the exchequer, Antony Barber, sharply rebuked Distillers, "astonished that they would make such a claim."

With that, the City lost all patience with one of its own. "Distillers were blundering round like a Glaswegian drunk," said one financial analyst, an impression that was heightened when a major chain of 260 stores announced they were boycotting all Distillers' products. Then, in the United States, "consumers' watchdog" Ralph Nader (decades before he became a presidential candidate) warned that a U.S. boycott would begin in a month if the British victims did not receive the equivalent of what U.S. victims David Diamond and Peggy McCarrick had received. With that, the total value of Distillers' stock dropped by $80 million in less than two weeks.

On January 3, 1973, Distillers announced it would be paying £2 million a year into the childrens' charitable trust for the next ten years, to be supervised by a number of independent medical and legal experts. The 341 families who had not been part of the 1968 agreement with the 62 that had, received the same inadequate amounts, totaling £6 million. Over the next seven years, Distillers would build a fund with £14 million, from which all 456 children would draw annually, depending on the extent of their damage.

Of course, Distillers expected all along that the actual costs to them would be negligible, since the company had liability insurance from Lloyd's of London—another vast economic force that all along had had a huge vested interest in seeing the thalidomide case remain unresolved. It was a surprise to the alcohol and pharmaceutical giant, then, when Lloyd's refused to pay out the total £10 million insurance claim Distillers submitted with which to pay the victims. Lloyd's of London refused to pay more than £3 million, on the grounds that Distillers had failed to conduct adequate tests, whereupon the pharmaceutical company backed down and settled out of court. That was not an argument they could afford to lose.

Six months later the High Court approved the arrangements, and with that, the goal set by the families had been reached. Needless to say there were further delays and many injustices, moral and probably legal, still to come. Future rises in cost of living had barely been accounted for, and most of the childrens' injuries still had to be assessed; with some birth defects it was not clear whether thalidomide was responsible. New families came forward periodically, some knowing full well that thalidomide could not have been responsible, whereas a few genuine victims didn't come forward for years. Still, after the 1973 agreement, the families in England who initially had fared among the worst in the world came out among the best.

But the country still had no idea how the catastrophe had begun, who was responsible, and, most important, whether it could happen again. Unless the public and Parliament learned the facts of the case, there could be no pressure for change. In fact, a lengthy debate in Commons was planned soon on the testing, labeling, and pricing of drugs—without anyone knowing anything about how the thalidomide epidemic came about. Since a handful of victims' arrangements were not fully resolved, the ban on the *Sunday Times* articles—and any other news story—was still in place, with a penalty of imprisonment for Evans or other editors who violated it.

In early 1973, three weeks after Distillers agreed to settle, the Court of Appeal unanimously set aside the ban on the *Sunday*

Times article. However, it allowed the attorney general the rare opportunity to appeal the case to the highest court in the land, the Law Lords.

In Britain, a panel of five Law Lords is what passes for a Supreme Court, chosen by rotation for each case from among the twelve Law Lords, who in turn are selected from within the House of Lords by the Lord Chancellor in a process that is completely and deliberately invisible, a fact that makes the highest court in the land barely distinguishable from a private club of elderly gentlemen. They are the only judges who do not wear white wigs or even judicial robes, as if aloof from the folly governing all the other courts of England. Now this ruling class, represented by Lord Reid, Lord Morris of Borth-Y-Gest, Lord Diplock, Lord Simon of Glaisdale, and Lord Cross of Chelsea, held the fate of the free press in their hands.

"It is ordered by the Lords Spiritual and Temporal in the Court of Parliament of Her Majesty the Queen. . . . " The *Sunday Times* never really had a chance. Their decision in July 1973 was unanimous: the Lords would not tolerate any threat whatsoever of "government by the media" and they would not allow the *Sunday Times* to publish the article or to give it to a member to read in Parliament or to convey it to any newspaper outside the UK to be published there. That was that.

There was an uproar. The most articulate expression of outrage came from Harold Wilson, still leader of the opposition Labour Party, in a letter to the *Times*. "Our debates and the ultimate outcome [of the settlement] were inspired and informed by the original article in the *Sunday Times*. The gravity of the House of Lords decision . . . lies in the fact that, had that decision been operative a year ago, there would have been no settlement at all for the victims of thalidomide." Wilson added that, if the Law Lords' interpretation of the law was correct, then the law of contempt of court had to be changed.

There would soon be grounds for doing just that—but not yet. The government had commissioned a panel report on the law in 1971. Three years later this report was presented by Lord Cameron, who firmly backed the paper. "What lent the *Sunday Times* campaign so much strength," he wrote, "was the fact that, in the eyes of many people, justice and the administration of the law in the thalidomide case were two very different things. We think there is great force in the argument. . . ."

But by 1975, when Labour won the general elections and Wilson became prime minister again, his outspoken enthusiasm for reform cooled considerably. He chose his battles with the Tories carefully, and contempt of court was not among them. So nothing could be done to inform the public and Parliament about grave flaws in the government's regulation of drugs that could, any day, result in another epidemic; nothing, until the ink dried on the last thalidomide settlement.

Evans had one last strategy that was fantastically daring by any standard. "Looking back," he said recently, "I can hardly believe I had all that nerve." Twenty years earlier Britain had ratified the European Commission's Convention for the Protection of Human Rights, which guarantees the right to free expression. The *Sunday Times* decided to take the question of contempt to foreign soil, south of Calais. This was an extraordinary measure, but Evans and many others felt it was so crucial to change the contempt law that nothing less would do—even after the government let the *Sunday Times* know that, since the last four childrens' cases were almost resolved, it was now prepared to allow the articles into print. But the government's right to gag the press in this and similar cases was an even larger issue to Evans, and it had to be resolved, albeit outside of British sovereign authority. The *Sunday Times*, having been subjected to so many accusations in court, was now putting British jurisprudence itself in the dock, in another country. The words "Bad form!" were much overused in Belgravia that week.

In December 1975, after the European Commission declared the case admissible (in itself quite remarkable), hearings were held, and that was that. Findings would be published in a year or two. But perhaps the European hearings—a black eye, internationally—had achieved the intended effect, because all of a sudden the attorney general backed down. In a four-minute hearing, Lord Widgery lifted the ban.

At last, on June 27, 1976, the *Sunday Times* published a six-page investigation, revealing for the first time that in 1958 teratogenicity was something other companies routinely screened new drugs for. But the long-suppressed article still did not include the critical information from Distillers' documents about animal tests, marketing decisions, and reports of nerve damage, among many other things. The *Sunday Times* Insight Team had had all this information for years, but could tell no one—not then, not ever. The injunction remained on these documents and their implications, because the manner in which the paper (and Phillips) had procured the documents constituted a breach of confidence.

Attempting to overcome this injunction in still another court, the *Sunday Times* cited an 1856 ruling: "There is no confidence as to the disclosure of inequity." The attempt failed. Despite the success in publishing significant parts of the story in 1976, the information most pertinent to the well-being of the British public simply could not be printed, anywhere in the world, without violating British law.

A year later, Harold Evans's fantastically daring strategy—the European gambit—unexpectedly paid off in spades. In July 1977 the European Commission on Human Rights reported their finding that the House of Lords had denied the editor and reporters of the *Sunday Times* their right to free speech. The European court "put its judgment in a way which appealed to me," Evans later wrote, "not so much on the right of the press to publish as the right of an individual to information which may affect his life, liberty, and happiness."

As a result, the British government would now have to defend its system and practice of law, specifically with regard to contempt of court. The last kind of justice one might ever have expected from the *Sunday Times* thalidomide case was poetic justice, but surely this was it: the new British government under Conservative prime minister, Margaret Thatcher, was now compelled to appear before a court of seven judges from seven nations to explain to the world the mysteries of English law.

But the European Commission's report included something even more extraordinary and utterly unexpected, which the British government could not suppress: the entire text of the forbidden *Sunday Times* article was included as an appendix to the report—the article that included the key documents obtained from Distillers, suppressed ever since they had been acquired ten years earlier.

Although the next day was a bank holiday, the paper sought an exception to the court's injunction that would allow it to print the European Commission's whole report, including the appendix. Had human rights in Britain eroded so far as to block a publication of the Human Rights Commission? At a special hearing in the judge's parlor at home, the exception was granted.

On July 31, 1977, the long-embattled article was finally published in England, including an embarrassing scientific blunder: the assertion by the Insight Team (which they later disavowed) that thalidomide's molecular structure should have suggested its teratogenicity to Distillers. Now the teenage victims' families could finally make some clear appraisal of what had been done to their lives, and who was responsible. The information Distillers had suppressed for sixteen years could at last enter the political dialogue.

Because of a deliberate work stoppage by spiteful members of a printers' union (who knew exactly how important this story was to Evans), half a million subscribers—one-third the total—didn't get their paper that day, and never saw the article.

No one was ever found guilty of negligence toward the 456 deformed children in Britain. No one was charged with a crime.

In their 1979 account of the ordeal, the *Sunday Times* Insight Team wrote: "Critics of the paper's thalidomide campaign assumed that there had been two equal parties to a rational contest, and that the legal processes would see that the truth came out and justice was done. That was palpably false. The parents of the damaged children were capable neither emotionally nor financially, after all those years, of fighting Distillers. Time put pressure on *them*, not on a corporation that had no handicapped children to worry about."

The notion that a corporation must act morally is relatively new. Indeed, Distillers' argument that its only obligation was to shareholders affirms the doctrine of the earliest joint-stock companies that produced the London of Charles Dickens, for better *and* for worse. Undoubtedly the joint-stock company provided an indispensable formula for capital formation, and so helped create the first large industries. Since the mid–twentieth century, much attention has been given to the responsibilities corporations have to their consumers and their communities, and indeed there have been dramatic changes in both attitude and action. Shareholders have generally favored companies with community outreach programs, charitable funds, and job-training scholarships. At the same time, the corporation's responsibility to its stockholders remains paramount, as it was with Distillers.

Less attention has been given to the responsibility of the shareholders of a company, when its management behaves unethically. As a matter of legal justice, there is none. But in the matter of moral justice, the shareholders of Distillers set a higher standard, as did the *Sunday Times*.

Reforming the FDA | 6

Thou among the leaves hast never known,
The weariness, the fever, and the fret,
Here, where men sit and hear each other groan.

—John Keats, *Ode to a Nightingale* (1818)

In the United States, unlike England, the thalidomide epidemic had a swift, significant, and lasting effect on the regulation of drugs, thanks in large part to the hearings begun two years earlier by Senator Estes Kefauver of Tennessee, initially concerned with the price of drugs. The hearings were slowly going nowhere, until the tragedy struck.

It was not really until the 1950s that prescribed medicines came of age as commercial products; by the end of the decade, there were 12,000 pharmaceutical manufacturing companies in the country and 56,000 retail drugstores. Many more medicines were synthetic, rather than extracted from a natural product, and there was enormous growth in the development and production of new "wonder drugs" as well.

But wonder drugs were expensive, and by the end of 1959 the American public was complaining bitterly. Small companies, for example, sold generic penicillin for less than $5 for 100 tablets, whereas giant pharmaceuticals sold the same product by their brand name for $12. Some of the difference in cost lay in research, most in advertising, and the rest in greed.

Many physicians prescribed only brand-name drugs. Welfare patients were the exception. Kefauver had observed that "the American Medical Association recommended that prescriptions be prescribed by generic name for welfare patients. But the same association failed to make such a recommendation for all other people." Doctors lacked faith in smaller companies that sometimes produced substandard drugs. This wasn't surprising; although one couldn't work as a pharmacist serving thousands of people without a license, one could legally manufacture prescription pharmaceuticals for millions. Unless a physician could feel confident that pharmaceutical plants were subject to effective inspection, it was unrealistic to expect her or him to prescribe generics. Eventually the Pharmaceutical Manufacturers Association proposed that Congress register pharmaceutical manufacturers and close down "bathtub" operations—fly-by-night companies that manufactured drugs without controls or personnel training.

But the main reason for doctors choosing brand names was their sustained promotion and advertising. Since physicians could not read 5,000 medical journals annually and run their busy practices, they tended to rely on advertisements to decide which medicines to prescribe for their patients. One congressman expressed shock even to hear it said that doctors could be easily swayed: "I think it is almost an insult to the medical profession to give the impression that they practice medicine from advertisements in medical journals." But surveys showed that ads in medical journals were the largest factor in their choice of medicines to prescribe, even though many were misleading, if not downright deceptive. One doctor who testified before Kefauver's committee noted there was often "a conflict between promotional material and scientific information appearing in the *same issue* of the *Journal of the American Medical Association*." Advertising financed fully half the activities of the AMA.

As one biographer of Kefauver described the battle for reform, "It was a fight for the 'little people' against powerful, well-financed special interests. It had the superficial support but surreptitious obstructionism of his own party's leaders." And the battle between the "little people" and the big drug companies came down to a duel between two old, senatorial alpha males, Kefauver and Everett Dirksen, the Senate's Republican minority leader from Illinois.

The Kefauver hearings in December 1959 were closely watched by an irate nation. In addition to members of the subcommittee and invited witnesses, the high-ceilinged caucus room of the Old Senate Office Building was crowded with several hundred spectators, including reporters, television camera crews, and drug industry representatives.

The first witness was the president of the Schering Corporation, which sold prednisone (prednisolone) at a wholesale markup of 7,079 percent. Schering didn't even manufacture the drug: they bought it from Upjohn and repackaged it for their profit. This was just what the angry American public was looking for, and the reporters in the room rushed to the wires. Next morning the *New York Times* trumpeted: "Senate Panel Cites Mark-ups on Drugs Ranging to 7,079%."

But Senator Dirksen challenged the attack, insisting that "the cost of production, research, selling, and distribution, administrative, and taxes [left] a profit of sixteen percent after taxes"; and he referred to the "glaring and misleading headlines and front-page stories. . . . Mr. Kefauver and his colleagues will have on their consciences many thousands of needless deaths and many millions of hours of avoidable suffering."

The protracted battle between these two craggy old senatorial bulls and Dirksen's efforts to provoke Kefauver, were beautifully described in the *New Republic:*

> Alas, [Dirksen] faced one of the most wooden antagonists in the Senate. Senator Kefauver's long, equine head, owl-like glasses, and invisible coonskin cap did not move. Dirksen raised his mellifluous voice and wagged the remnants of his curly hair. Kefau-

> ver sat like an absent-minded doorpost. Again and again the emotional Dirksen set lance and charged—the protector of doctors and druggists. Milder and more limp came Kefauver's languid response, always introduced with an infuriating pause and inquiry if the distinguished Senator from Illinois were finished. Assured by the fuming Dirksen, Kefauver deprecatingly repeated his damning statistic, his voice never changing, and continuing with the irritation of a leaky faucet. Dirksen, who has the face of a lost angel, writhed in distaste—like an alcoholic offered a bowl of warm milk. . . . Finally, Dirksen stopped trying to fire Roman candles into this pile of damp sawdust. The two veterans, who had hitherto hardly exchanged glances, softly turned at the end and grinned.

The large pharmaceutical companies, of course, not only supported Dirksen: drug industry lobbyists wrote some of his speeches, which he delivered without hesitation, presenting the pharmaceutical companies' arguments as his own. For example, he declared that companies averaged about one new product for every $5 million they invested in research and that, in 1958, the industry tested 114,600 substances to produce just forty marketable drugs.

An editorial in the *Chicago Sun-Times* responded to the hearings as well, noting that most drugs prescribed today were not even in existence ten years ago. "It is ironic that Senator Kefauver protests the cost of drugs for elderly people, although some of them are still alive because of the new drugs." The pharmaceutical giants had many friends in high places. Even an essay in the *New York Herald-Tribune* sympathetic to Kefauver admitted that "the Senator has often allowed his own zeal to outrun discretion." The debate dragged on.

Despite all the attention paid to the ongoing hearings, when news first came of the thalidomide tragedy in Europe in November 1961, it was almost completely ignored, like so many news stories from abroad.

In his message to the legislature in January 1962, President Kennedy pressed for action to reduce the high cost of prescription drugs. In March he followed up by telling Congress that "The physician and consumer should have the assurance that any drug or therapeutic device on the market today is safe *and* effective for its intended use. . . . There is a need for new legislative authority to advance and protect the interests of consumers in the marketing of drugs." Kefauver's latest bill, said Kennedy, was just what was needed.

Congresswoman Leonor Sullivan (there were no congresspersons yet) of Missouri observed that "drug legislation then pending at the Capitol was given a push by President Kennedy . . . but the push didn't go very far. There were warnings that something could go drastically wrong if more controls weren't placed over the mass of new drugs, but nobody around the Capitol had heard of thalidomide." For the first six months of the unfolding crisis in Europe, the public in the United States was only dimly aware of it.

The Democrats' solution to problems in the drug industry, not surprisingly, was for the federal government to control it more tightly. As well, patent law would be changed so that exclusive market rights would drop from seventeen years to three, inviting competition sooner, and the small companies would be controlled by tighter inspection laws. Physicians would be encouraged to write prescriptions by generic rather than by trade name and labeling laws would be reformed to facilitate such a change. Senator Kefauver noted that "free competition has been hampered by patent monopoly control." The underlying problem was that, as Kefauver put it, "He who buys the medicines does not order them, and he who orders them does not buy them."

The Republicans agreed that prices were too high, but argued that drug prices were not unique in this regard and cited examples where drug prices were not increasing at a rate greater than many other goods. They argued that the U.S. patent system drove the economy: without patent protection there would be no incentive for companies to invest in the discovery of new remedies. As a result of our current system, bragged the free-enterprise devotees, the United States had discovered twice as many new drugs since World War II as all the European countries combined. The USSR, where there was no patent system, had invented none. "The point at issue," observed Senator Kenneth Keating of New York, "was whether regimentation of this industry in the manner originally proposed by some would go so far in stifling research that unwittingly more harm than good would be done."

When the sessions reconvened on June 11, 1962, a heated discussion erupted on the Senate floor. Kefauver had been blindsided and he was angry. "Most of the drug manufacturing industry and its acolytes have been punching away for some time. Today they swung a haymaker and just about knocked this bill right out of the ring. Much to my amazement I discovered that there had been a secret meeting between representatives of Health, Education, and Welfare and staff members of the Judiciary Committee and Antitrust Subcommittee of which I knew nothing. The bill which now remains is a mere shadow of the one approved by the Antitrust and Monopoly Subcommittee."

Senator James Eastland of Mississippi took the blame. "I accept full responsibility for the alleged secret meeting. It was my obligation to do what I think was needed, to get a realistic drug program. I did not call in my friend from Tennessee because I thought it would be a futile act. I did not think he would make any agreement with respect to anything. I do not think the iron heel of the U.S. Government should be placed on any company which wishes to manufacture drugs."

It was in this atmosphere of frustration, uncertainty, and downright animosity that the thalidomide story finally caught the country's attention.

During subcommittee hearings in mid-May, FDA commissioner George P. Larrick referred to the "German experience" without specifically naming the drug. Then, Dr. Helen Taussig testified before the subcommittee about her personal study in Europe, concluding, ". . . had this drug been invented in this country, I believe it would have passed our present laws, as it is only under special circumstances that tests of pregnant animals are requested." But it wasn't until midsummer that the press finally grasped the full implications of the thalidomide tragedy to the United States, when the *Washington Post* carried the story about Dr. Frances Kelsey, catapulting her into the spotlight.

Congresswoman Sullivan later allowed that, without the *Post* story, "I imagine we would have had no bill at all this year. This article was the trigger. It came at a time when the best we could hope for was a watered-down bill which may or may not have passed the Senate, with no House action of any kind scheduled."

It was exactly what Kefauver needed to build a head of steam; three days later, on the Senate floor, he announced that Kelsey "deserves a gold medal for distinguished Federal civilian service," and sent a letter to the president recommending her for that award.

The Harris Committee on Interstate and Foreign Commerce, as well as the Kefauver Subcommittee and the Humphrey Subcommittee on Reorganization and International Organizations and Foreign Commerce all began to investigate. After Congressman John Fogarty of Rhode Island noted that the nation was shocked by the details of the thalidomide tragedy, he got down to business. "Although our country's drug regulations have been praised as the most comprehensive standards of any country in the world, it has become evident that they are not sufficient for our needs. The recent and highly regrettable thalidomide tragedy has left no doubt that the need for such strengthening exists."

Senator Kefauver must have worked hard not to seem to gloat. "During our investigatory hearings the pharmaceutical manufacturers testified that everything was all right as it was. They were against everything we proposed." But after they had learned about thalidomide, "they accepted about three-fourths of the provisions of the bill. In contrast, representatives of the American Medical Association stated that in this entire omnibus bill there was nothing which they could support . . . [even though] in my opinion, ninety percent of the doctors favor the bill. Even the conservative *New England Journal of Medicine* carried six editorials endorsing and recommending the passage of most of the bill."

When Senator Philip Hart of Michigan was asked by a member of the press whether he believed that society must face a disaster before taking action on a domestic issue, he answered, "If it had not been for the thalidomide incident, this bill would never have gotten out of the committee." Congressman Peter Rodino of New

Jersey agreed wistfully. "I do not think we need to be particularly proud that it took an international catastrophe to make us realize that the first thing with drugs is safety."

Alarmed by Richardson-Merrell's wide distribution of thalidomide in the United States without approval, the Senate committees wanted to know who was qualified to conduct so-called "clinical trials." Deputy commissioner of the FDA John L. Harvey answered frankly, "It may be argued that every doctor is a qualified investigator." Senator Jacob Javits of New York agreed: "Any physician who signs a document stating that he is an expert is thereby deemed to be an expert."

The Senate committee wanted to know the opinion of the nation's new hero, Dr. Kelsey. "Unfortunately," she said, "there is a scarcity of persons with training in drug evaluation. Many of the investigators who supply the information to the companies have no experience in this line. The reports show it. In many instances they are in the form of testimonials."

Congressman John Bennett of Michigan stated the obvious. "It would seem elementary that an investigator using an experimental drug on patients should be required to establish and maintain records on his experiment." Only 16 percent of the physicians involved in Richardson-Merrell's clinical trials had reported any information at all back to the company. Six physicians who were "testing" the drug donated supplies of thalidomide to religious groups for charitable distribution overseas. In at least one case, six bottles of thalidomide donated to a religious group were traded to a hospital pharmacy for other drugs. With millions of tablets being shipped and traded, the likelihood is great that many victims around the world and their families never knew what hit them—and could never find out.

Perhaps the most shocking revelation about the "clinical trials" of thalidomide was that, in Senator John Carroll's words, there was "no evidence that the doctor receiving the drug told the patient that he was to be used for experimental purposes. I believe firmly every human being has a right to know whether he is being treated with experimental medicine," though, in accord with the custom of the times, Carroll had to add a proviso. "I realize there may be cases in which a doctor cannot inform his patient of the treatment, because the patient may be suffering from a serious case of cancer, and the doctor has not informed the patient of his illness." This "protectiveness" may well seem strange today, but Senator Eastland among others was quick to defend it, "in view of the fact that if the patient knew of his condition, the emotional reaction would cause him greater harm." (In its final form, the Kefauver bill contained the following compromise language: "except where obtaining such consent would not be feasible, or in the professional judgment of the investigator would be contrary to the best interest of the patient"; that exception was removed by further legislation in 1966.)

It did not escape the senators' attention that the informed consent of a patient involved in a clinical trial was a crucial element in what became known as the "ten commandments" of human research: the Codes of Medical Ethics that had evolved out of the trials of "The Twenty-Three" Nazi doctors. In drug tests (and other procedures), "the voluntary consent of the human subject is absolutely essential," the U.S.-led Nuremberg Military Tribunal had concluded in 1947. But the same requirement did not exist under U.S. law when Richardson-Merrell sent out 2.5 million pills for promotion; if it had, the company could have faced criminal charges for their recklessness toward Peggy McCarrick, David Diamond, and the nine other U.S. victims.

President Kennedy wrote a letter to Senator Eastland in August 1962 with seven recommended amendments to the pending drug legislation, requiring that: the FDA be informed in advance about clinical investigations, which had to be executed by named, qualified investigators; and, if a substantial doubt developed as to the safety of the drug the trial would be halted by the FDA. Hardly an unreasonable burden, one would think, for the privilege of putting people to some unknown degree of risk. Four days later, to give the bill another timely push, he awarded the gold medal to Dr. Kelsey in the White House Rose Garden.

Two weeks later, on August 23, 1962, Senate Bill 1552 was passed by a vote of seventy-eight to zero, establishing FDA regulations that, with very few changes, are still the law of the land. That day, Senator Dirksen did not vote. Two months later, after the Senate bill was reconciled with the similar House bill sponsored by Congressman Oren Harris of Arkansas, President Kennedy signed into law the Kefauver-Harris drug amendments to the Food, Drug, and Cosmetic Act.

Senator Kefauver, with a nod to us today, allowed that "those in the future who attempt to study the legislative history of this measure as it passed through its various stages may be forgiven if they become somewhat confused." The bill that Senator Kefauver first introduced in 1960 was intended to decrease the cost of drugs; the bill signed into law unanimously in the wake of the thalidomide epidemic had been defanged of its price-cutting teeth and was replaced instead by crucial regulatory safeguards, which, in and of themselves, raised the prices of drugs dramatically over the years to come.

Children's Voices, Strong and Clear

7

All happy families resemble one another, but
each unhappy family is unhappy in its own way.

—Leo Tolstoy, *Anna Karenina* (1875)

In clumsy and cruel ways, medical science continually tried to make amends to the children who had been mutilated by medication. Around the world, many of the 5,000 or so who survived infancy without being condemned to an institution spent their childhoods as medical specimens. Even the most healthy kids saw many doctors. They were analyzed, measured, and blood-tested repeatedly and meticulously; most of all they were X-rayed continuously as their deformed limbs developed. A great deal of surgery was performed, some life-critical, some purely cosmetic. The twenty or so British children born without ears went through as many as eighteen surgeries to produce something that resembled ears, though they remained completely deaf. In some cases, surgery was performed to remove the little stubs of limbs; the stunned parents may have agreed to these procedures out of despair, or even aesthetic revulsion. According to Dr. Newman, "With everything that was extra, not quite right, or sticking out, the tendency was to say: 'Off with it!' No one saw how these remnants of limbs could be invaluable in later life for operating switches or tools, and it was the parents who realized this before the doctors." The results left some victims quadruple amputees.

The Utopian prosthetics custom-designed for the deformities caused by Utopian medicine were crude. British thalidomider Kevin Donnellon, interviewed by BBC, reflected on the device built for him when he was a young man: a plastic armor suit, using the cutting-edge technology of the 1960s to provide him with mechanical arms. "I looked totally ridiculous in them. I looked like a robot, but they were totally, totally unfunctional. I mean I couldn't do anything in them really. These are arms which are activated by gas cylinders which, you move your shoulder then these claws move. I mean they were totally useless, I couldn't pick anything up with them. Very, very impractical and they were extremely heavy, and I felt really like caged in. The gas wouldn't last that long, and sometimes you'd get it halfway to your mouth and then the gas would run out. The legs in some ways were worse, because they were far more dangerous—extremely heavy, you know. The design hadn't been changed since the First World War, you know; they were made of cast-iron and plaster. They were extremely heavy, and also, I didn't have the balance with having no arms and, you know, you fall to the ground like that. It was the constant fear, you know, any second I expected to fall forwards or backwards, and, nine times out of ten, I'd just fall on me face or have stitches in the back of me head."

The first mechanical arms for the children were designed in Germany. They came to be called "pat-a-cake" hands, because that was all they were good for, and they were not very good at that. The fancy models that came later were not much better, and involved the children and their parents in hundreds of hours of frustration before the contrivances were finally given up. "The experience has benefited many of today's handicapped children," says Vaughan, "but it has not taken away from the damage done to the families at that time. Those children had to go through psychological trauma and considerable pain just to satisfy some community conscience that the doctors had tried to do something to put the tragedy right."

In Canada alone, three sets of twins are thalidomiders. It is important to note that, even when the twins are identical, the defects of the twins are not, as one would expect if their deformities were the result of genetic mutation, instead of chemical mutilation. The mother of two boys was anesthetized immediately after their birth, just so that she wouldn't see them. One of those twins later wrote: "I was born with one leg and a stump with a foot. The doctors amputated the foot in order for me to be able to wear an artificial leg. I had about seventeen operations in all. . . . As I had a twin with similar disabilities, it made it easier growing up, because we helped each other out. We were able to help pick each other up, and we didn't feel so alone with the problems we faced in everyday life."

Much of Randy Warren's early life at the Shriner's Hospital in Montréal involved the custom development of artificial limbs and the skills to use them. Randy spent hours each day strapped into legs that flailed helplessly, naked, parading in front of the hospital staff. When he remembers his adolescence, that is what he recalls most distinctly.

The artificial limbs were among the greatest sources of misery for almost all the victims, and consumed much of their young lives while remaining, for all but a few, completely useless. It is hard to exaggerate the psychological damage inflicted by these years of futile effort. It was a hideous and almost universal mistake that undoubtedly diminished their lives and haunted some victims as their first failures—theirs because, of course, it *couldn't* be the experts' fault. In Canada, artificial limbs were provided almost forcibly by the government: in part, this exercise provided research grants to prosthetics centers that served military veterans who were amputees. But it was also a means of expiating medicine's sins to these children. So although some of the families remained destitute, almost every one of Canada's 115 victims who were limb-deprived had to be fitted with these utterly ineffective prosthetics.

Much later Randy learned from the medical consultant to the War Amputees of Canada, Dr. Gustave Gingras, that it is almost impossible for someone who has *never* had limbs to master these artificial devices. The fundamental nerve pathways between mind and body simply are not there; so, for example, a thalidomide victim would never experience pain in a "phantom limb," as amputees do.

Much of Randy's childhood was spent showing his naked, distorted, truncated body to strangers. Doctors from all fields of medicine came to observe him, many to treat him, others for the righteous purpose of identifying other victims, and some from the same base curiosity that propels people to sideshows. Even doctors recoiled from him at first before recovering their professional demeanor.

Randy developed a jaded perspective of these detached scientists, which soon turned to a cynicism that would not quickly fade. Almost like a traveling carney, he came to see his visitors as "marks," ripe for manipulation. He had told the same story so many, many times—the one that began, "My mother was given some medicine . . . "—that it became a riff in its own right, one that he could sometimes recite without any apparent emotion, or, with the most effective emotion to get what he wanted.

Other kids came to Shriner's with club feet, spina bifida, and cerebral palsy, and they left not long after surgery; so there was a pecking order determined by how long you'd been a patient, how many Christmases you had spent there, and how rarely your parents came and visited and left you crying in the lobby. That made Randy king of the hill for years in a row, although he did return home periodically, between surgeries. The hospital was partly staffed by nuns, who brought their own formulae for discipline, but for the most part there was regularity to life and an attempt to imitate the natural life of young adolescents. Pranks by the kids were tolerated and even perhaps encouraged. Sometimes Randy and some other boys would "sneak out for a cigarette" they had cadged from one of the staff. But, of course, a half-dozen disabled boys with artificial limbs and wheelchairs aren't very good at sneaking anywhere, and despite hours of scouting and transporting, Randy often got left at the bottom of some stairway.

At sixteen he had to leave Shriner's Hospital, which only cared for children, and return to his parents in London, Ontario. He wasn't allowed to use a wheelchair except on special occasions, but instead was ordered by doctors and parents to get around awkwardly, dangerously, balancing on his hated, painful artificial legs, knowing all along what we all know instinctively: Everyone finds their way to the cookie jar if they want a cookie bad enough. That's a well-known motto among the handicapped.

Since his parents wouldn't allow him to leave home without the legs, as the doctors insisted, Randy stayed in the house for two years. Never having lived there full time before, that alone was a shock. But it created opportunities for emotional exchanges that had never happened before, and that everyone needed.

One afternoon when Randy was alone with his mother, Anneliese, the question "*Why?*" erupted against his will, the question he'd been choking on since he was nine years old. "*Why did you take that medicine?*"

His mother gasped. Then she sat down beside him and looked in his eyes. "A doctor gave me medicine and he told me I should take it. I was a young woman with a first pregnancy, and I trusted the doctor. So I have never suffered one moment of guilt. The responsibility is not mine, Randy. It is not mine." His love for his mother and his overriding compassion for all she had suffered from the day he was born (when her own life was threatened because the obstetrician could not find his legs), and the limitless love she had given him—all that overcame forever any recrimination he might have felt. He would not waste his life in anger. Instead, he continued learning to live life on life's terms, without ignoring his own requirements, first among which was respect from others. Instead of calling himself a victim, Randy started calling himself a "thalidomider," to keep the attention on the drug he despised. As his education proceeded, he came to know a few other victims, and learned the fates of many more, much worse off than himself. In Brazil, he knew, most thalidomiders his age had been begging in the streets all their lives. Much later he learned that some became prostitutes, and that there were terrible, terrible photos of them in circulation.

As he looked about his immediate world, Randy began to feel less bitter about his own situation. The medical world that had tortured him had also saved his life. And he loved his sculpted hands, each with four fingers, each with opposing thumbs created by removing one finger completely and repositioning it—the surgeons' best contribution to his adult life, and he was very proud of them. Now he had to decide what to make of his life, given what he had to work with. He received a pension from the German government (minus the crucial health benefits he would receive if he lived there), and there had also been the settlement with Grünenthal, eventually, after the trial at Alsdorf. He had parents who loved him dearly, as well as a younger sister and brother. Randy was just starting to teach wheelchair basketball. He grew a mustache.

But all tranquillity vanished when he learned that surgeons were planning to cut off both his feet.

That was the major turning point of his young life. All the pain and anger, the lifelong frustration and relentless humiliation, exploded. Disgusted though he was with his own body, his desperate need for physical integrity overcame his docile nature, and he fought for all he was worth. He had already mutilated his thighs with a razor blade, so that they couldn't force him into the artificial legs; now he threatened suicide and refused to eat. His parents kept insisting, reminding him of his progress from the other surgeries. With these *last* two operations, the doctors said, the artificial legs would finally work the way they were supposed to.

Some parents discovered much later that thalidomide children, who might survive without obvious deformities, were prone to yet another unthinkable tragedy: autism. Having gradually come to hope for some modicum of well-being for their child's life, these parents discovered a few years later that their child was also imprisoned in permanent emotional and intellectual isolation. The developmental connection between thalidomide and autism has been established only recently. Dr. Patricia M. Rodier has reported on the little-understood biological basis of autism, which normally affects 16 out of every 10,000 babies. That rate is thirty times higher in thalidomide victims; roughly 5 percent of those who survived infancy turned out to be autistic. She noted that the drug apparently produced this result when the mother ingested thalidomide between the twentieth and twenty-fourth day after conception—well before most of these women even knew they were pregnant. The fact that many of the autistic children whose mothers had taken thalidomide possessed normal limbs led Rodier to conclude that: a.) thalidomide affected the development of the cranial nerves, affecting the eyes and ears and also the brain stem (which develops before the limbs), thereby causing autism; b.) non-thalidomide-induced autism develops much earlier than possibly imagined; and c.) apparently a specific gene, known as *Hoxa1*, is the most likely candidate to explain genetic autism, in part because its active phase is limited to the period in which thalidomide causes damage to the brain stem.

It is easier and more comforting, of course, to focus upon the success stories. Tony Melendez is certainly one of those. He was born in Rivas, Nicaragua, in early 1962, the second son of a fun-loving couple. Tony's father, who worked in a sugar cane refinery, often played guitar for his friends and neighbors, and hunted for crocodiles and boa constrictors. Tony's mother was prescribed thalidomide by her own uncle: she took the new German sedative for two weeks before she even knew she was pregnant.

Tony was born with no arms, eleven toes, and one club foot bent backward toward his leg. As he later chronicled in his book, *A Gift of Hope* (written with Mel White), his parents were, of course, devastated. But from the beginning of his life, his grandmother insisted that God knew what He was doing when He sent this baby to them. With that unshakable religious conviction overriding all other considerations, his parents set out to give him a full and rewarding life, even though Tony often heard others describe its beginning as "his tragic birth."

Among Rivas's 9,000 inhabitants, word spread of "the Melendez monster," and children and passersby began flocking regularly in front of the house—not with malice, only curiosity. "After hearing Mom's stories about how hard people worked to see me," wrote Tony, "it's easier to uderstand why P. T. Barnum was such a success with his traveling circus of midgets, two-headed calves, and fat ladies who grew beards." In the endemic poverty of rural Nicaragua at the time, the helpless, misshapen baby could be expected to end his years as a beggar in the streets.

When he was four months old, Tony had the first of a series of surgeries on his club foot in Managua, three hours from Rivas along washed-out roads. When the family won $5,000 in a lottery his extra toe was surgically removed. Then his father packed ten members of the family into an old Chevrolet and set out for Los Angeles, 5,000 miles away. After a grueling, seventeen-day trip, they all settled in with family in east L.A. and Tony's father got a job in the garment industry and began to work toward naturalization: only as a U.S. citizen could the family receive the financial assistance they would need to pay for Tony's extensive surgeries. Before that happened, the March of Dimes stepped in and volunteered to pay medical bills for the child.

As soon as Tony had full use of both feet he began to walk, although one leg was longer than the other. He also developed special skills. "With my toes I could hold a hand of cards, tune the radio, grip a hammer, pound round wooden pegs into holes, and even pinch my brother's rear end." He was fitted with one artificial arm. "Learning to use it was a nightmare. A thin steel cable hung down from the shoulder harness and was attached to my right leg. When I shrugged or stretched my shoulder correctly, the cable would tighten and the arm would move up and down." Tony gave up this prosthetic device, but eventually wore another, which he also hated, for six long years: "I lost or hid that arm on every possible occasion."

In the meantime, Tony continued to develop his skills with his feet. He won a snooker championship at his orthopedic school in Chino, California, where the family settled, and he could ride a skateboard, traveling all around the community. He swam like a fish under water, and, having been accepted at the "regular" public high school, he made it onto the varsity soccer team. He got a car specially outfitted so that he could drive with just his feet. In short, Tony had developed a new approach to life, summed up in the words, "Who needs arms?"

Through his school choir and his church, Tony became a talented singer. He also learned to play his father's guitar (in open tuning) while he sang, strumming with a guitar pick gripped between his toes. For a few months he played and sang for quarters on a street corner in Laguna Beach. And just about the time he began to feel like a beggar, he received a very special invitation.

In a performance broadcast around the world, Tony Melendez sang and played for Pope John Paul II in 1987 at the Universal Amphitheater in Los Angeles, accompanied by a full symphony orchestra. He was twenty-five years old. A few years earlier, he had been refused admission at a seminary, to join the priesthood; the reason, he had been told, was that he lacked the thumb and forefinger with which to serve the Eucharist. Now, as the crowd cheered Tony, in an unrehearsed moment the Holy Father stood at his feet and, kissing his cheeks, said, "Tony, you are truly a courageous young man. You are giving hope to all of us. My wish to you is to continue giving this hope to all the people."

In the last dozen years, Tony has started his own ministry and his own record company—Toe Jam Productions. He has performed in forty-nine states and fourteen countries, as well as on every major talk show, and sang the national anthem at the 1989 World Series. The following year he married and settled in Dallas with his wife and daughter—having transcended more obstacles than most of us are capable of imagining.

Some of the young victims grew up to be doctors, others became scholars, and one German boy fully achieved his farfetched dream of becoming a world-renowned tenor. But these people are a tiny minority. It required a succession of enlightened guides to appear at just the right moment in their lives—parents, doctors, nurses, therapists, and teachers—to lift each young victim into adulthood, when often it was all they could do to manage the physical lifting required. Enlightened guides are always in short supply. Financial resources were a crucial factor in how the families and the children turned out, but they were certainly not the determining factor. No one could ever say what made the difference between the upbringings that made for "success" and the specifics of the damage in each individual made them impossible to compare. Still, it can be said without surprise that the children thrived best in families capable of unconditional love, as all children do. But for them, empathy was more than an emotion: if the parents could see the world from their child's perspective and use that knowledge to sort out which advice from the "experts" was good and which was not, their children could learn by themselves how to cope in the world with whatever equipment they had.

Early on, these children learned a lesson most of us take much longer to absorb: that you have to play the hand you are dealt.

LAZARUS RISES | 8

A certain beggar named Lazarus was laid at his gate, full of sores.

—LUKE, 16:20

WHEN LAZARUS THE LEPER WAS IMMORTALIZED in Jesus' parable, leprosy was already endemic throughout the Holy Land. One of the most dreaded diseases in human history, its name directly conjures up images of biblical times: figures crouching in the shadows, clad in rags to hide the flesh falling off their bodies, hissing "Unclean! . . . Unclean!" to warn away passersby. But Hansen's disease—the textbook name for leprosy—did not end with biblical times; hundreds of thousands of people around the world are currently affected by the disease, which is caused by a subcellular mycobacterium, M. *leprae*.

The Jerusalem Hospital for Hansen's disease was established by nineteenth-century German Protestants, as commemorated by the promise above its entrance: "Jesus Hilfe" ("Jesus Helps"). Built by the well-known architect and theologian Conrad Schick at 17 Rehov Marcus in the part of Jerusalem known as the "Forbidden

City," the striking, symmetrical building had been the home for as many as twenty patients at one time since before the turn of the century. The hospital's director through most of the second half of the twentieth century was not a German Protestant, but a middle-European Jew: Dr. Jacob Sheskin, who had survived the Holocaust by emigrating to Venezuela, where he specialized in treating leprosy patients until he resettled in Jerusalem in 1956.

There is no cure for leprosy, although antibiotics called sulfones are effective at controlling many of the symptoms. But sulfones have no effect whatsoever on a severe, inflammatory, and agonizing complication of leprosy known as ENL (Erythema Nodosum Laprosum), which occurs in about 60 percent of patients with the most severe form of the disease. The symptoms are large, persistent, weeping boils all over the body. Victims of ENL experience so much pain that, before the late 1960s, they were given morphine several times a day and Novocain was injected around the large, boil-like sores. They also had crippling joint pain, headaches, and abdominal pains, as well as severe inflammation of the eyes that led to blindness. Since patients could neither sleep nor eat, they became emaciated and often died from wasting.

In 1964, a critically ill patient with advanced ENL was referred to Dr. Sheskin by the University Hospital of Marseilles. The man had been bedridden for nineteen months, and by this time, on the verge of death, he was almost deranged from the unremitting pain that had denied him sleep for weeks; doctors in France had tried every existing sedative, but nothing he had been given helped for more than an hour.

Searching through the small hospital's infirmary for something that might work, Sheskin found one bottle of twenty thalidomide tablets. By this time, of course, the drug had been withdrawn from the market; but he knew that certain mental patients, whom no other sleep aid had helped, had been effectively treated with thalidomide. With nothing to lose in this case, the patient was given two thalidomide tablets as a last resort.

He slept soundly for twenty hours, and, upon waking, felt well enough to get out of bed without assistance. After two more pills, his pain disappeared entirely and his sores began to heal. When treatment was stopped, the symptoms returned, and when it was resumed, his condition again improved dramatically. Six other patients in the hospital were then treated, with similar dramatic results.

Dr. Sheskin was also a member of the teaching staff at Hebrew University, and the more he discussed these data with his colleagues, the more wildly unlikely it seemed. Thalidomide cures leprosy? That sounded like some awful joke. In fact, the whole notion that thalidomide could be beneficial to anyone seemed repellent: immoral, *prima facie*, like planning to make good use out of Mengele's statistics. So Sheskin felt he needed more evidence before he could conclude that thalidomide was so effective for ENL. Since thalidomide had been taken off the market in Israel, Dr. Sheskin traveled to Venezuela, where leprosy was still endemic, where thalidomide was still available, and where he had spent the war years. There he conducted clinical tests with patients whom he had treated in the past and for whom he had extensive medical records. Because the whole experiment seemed so wildly unlikely, it was extremely important (and proof of Sheskin's scientific diligence) that his tests were conducted meticulously, in a double-blind study with a control group. Half the patients were given the test drug and half were given a placebo, or sugar pill; neither the patient receiving treatment nor the prescribing physician knew which patients were given the test drug.

Of 173 leprosy patients studied in Venezuela, 92 percent of those receiving thalidomide were completely relieved of their symptoms. Dr. Sheskin published his findings in the journal *Clinical Pharmacology and Therapeutics* in 1965. A much larger follow-up study by the World Health Organization—involving 4,552 ENL patients in sixty-two countries—demonstrated, amazingly, that fully 99 percent of patients with this agonizing condition showed improvement within the first twenty-four to forty-eight hours of treatment, and total remission during the second week of treatment, the inflammation reduced and the boils vanished. For leprosy patients, the "monster drug" was indeed a wonder drug.

As a result of Dr. Sheskin's unintended discovery, thalidomide arose from the ashes of infamy. Thirty-five years later, it still remains the drug of need in treating ENL: nothing else comes close.

For that reason, in most countries with severe leprosy infestations, thalidomide has been available, one way or another, since 1965. Without it, hundreds of thousands of ENL patients in scores of countries would be in hospitals today, suffering unbearable pain and waiting for the blessing of death; with thalidomide, only 10 percent are hospitalized, and all the rest are treated as outpatients.

Because of Dr. Sheskin's discovery, 90 percent of the leprosy hospitals around the world were shut down. He was awarded a gold medal from the Laboratorio de Investigaciones Leprologicas, Argentina, in 1969 for his enormous contribution in relieving the suffering of hundreds of thousands of lepers, and then the French Gold Medal in Science from the Société d'Encouragement au Progrès in 1975; in 1997, two years before his death, he was named Worthy Son of Jerusalem (Yakeer Yeroshalayim).

The largest problem Dr. Sheskin had from the beginning was in obtaining the drug. There was, of course, only one place to go: the Chemie Grünenthal patent on thalidomide would not expire until the mid-1970s, and the company had a considerable stockpile left over from its sudden withdrawal from the market. Heinrich Mückter remained associated with the family business: he had been experimenting with the compound, through the 1960s, with cancer cells in vitro, and published several papers on the subject.

To obtain thalidomide, the European Jew who had spent the war years healing lepers in Venezuela sought out the former director of virus research for the Nazi killing machine in Poland—at about the time Mückter was preparing his criminal defense for maiming babies. What kind of relationship the two men had is not known, but as late as 1983, these two unlikely allies collaborated on a research paper about the drug. It was probably galling to the meticulous Sheskin to connect his name with such a notoriously careless scientist. Throughout this time Grünenthal continued to provide small quantities of the drug for compassionate use, thanks to a program set up in 1975 by the U.S. Public Health Service. From this Mückter extracted some measure of vindication for himself and the company.

After Sheskin's research in Venezuela, where leprosy is prevalent, small companies in South America began producing the drug in irregular batches, and of dubious purity and stability. Of course, despite its unforeseen success, thalidomide's dark consequences did not disappear. Although precautions were taken in the treatment of leprosy patients (for example, in Israel thalidomide was only given to female lepers who had been "rendered infertile"), victims of the drug continued to be born.

In December 1996, E. E. Castilla and his colleagues from the Latin American Collaborative Study of Congenital Malformations published the first review ever on the incidence of thalidomide-induced birth defects in South America. (That in itself is quite amazing, considering that thalidomide has been continuously present in several South American countries since its introduction in the late 1950s and all the more surprising given that the collaborative study began in 1967 as a result of the thalidomide disaster.) In all the years that the collaborative study had been under way, it had failed to identify thalidomide as a teratogen in South America. That is probably because the study covered a small, urban population and leprosy is mostly a rural disease, transmitted in the soil and, because thalidomide was already there when the study was initiated, thereby contributed to the baseline statistics. Brazil, with the greatest number of reported leprosy cases in the Americas, also has the largest number of thalidomide syndrome reports. As soon as Castilla's group looked for the drug's victims among the leper population, they found them. According to their report, thirty-four thalidomide babies have been born in rural South America—some as recently as 1995; another report puts the number at forty-seven. In all likelihood, the true number is much higher.

With the drug's success in treating leprosy, Dr. Sheskin began to consider the possibility that thalidomide might be effective for other inflammatory diseases, including AIDS, and he proposed the idea to colleagues in several universities, since he was too ill to conduct such studies himself. The torch would pass to the next generation of researchers. The next question was, "*How* does thalidomide combat the symptoms of leprosy?"

Dr. Gilla Kaplan became a research immunologist through a path that wove together many of the threads in thalidomide's story, long before she became the world's leading researcher with the drug. A woman of poise, warmth, and intellectual gravitas that is almost palpable, she was born in Johannesburg, South Africa, the daughter of both a doctor and a surgeon. Her mother, a primary health care doctor, worked with patients in the shantytowns at the edge of Johannesburg, where tuberculosis was rife. In 1952 the family emigrated to Israel, where Gilla's father, a plastic surgeon, occasionally performed reconstructive surgery, it so happens, on leprosy patients at Dr. Sheskin's clinic.

After completing her bachelor of science degree in biology, Kaplan went to the United Kingdom. She was in London in September 1972, when the *Sunday Times* published its groundbreaking article about the plight of thalidomide victims; Kaplan remembers actively supporting the boycott of Distillers that autumn, when, all around London, protesters put up posters of a whisky bottle marked "Thalidomide." She continued her studies in Tromsø, Norway, before coming to the United States, where she has been, for many years, associate professor at the Rockefeller University's Laboratory of Cellular Physiology and Immunology. As an immunologist she examines, in test tubes (*in vitro*) and in patients (*in vivo*), the interactions of potentially beneficial drugs with various immune cells.

Her job, then, as a disciplined scientist, is to find the right questions to ask, the right tests to perform, and then to eliminate from her interpretation of the data any expectations, assumptions, biases, or hopes she might have, to see the significance of the results with objective clarity. That clarity can make the difference between finding a cure for an incurable disease and raising false hopes for millions.

"I settled upon immunology early on," she recounted during a recent conversation in her office, which is not much larger than a newsstand, "and I became intrigued with mycobacterial diseases, perhaps because they were familiar to me from my mother's work in communities with tuberculosis and my father's work with leprosy patients."

Both diseases are caused by bacteria—intracellular parasites—that can invade specific immune system cells—the monocytes and macrophages. These cells have been the focus of Kaplan's work, as well as that of her senior research associate, Dr. Victoria Freedman. It took them years, along with other researchers around the world, to determine how and why thalidomide benefited leprosy patients, but when the answer came, it helped produce a revolution in a critical area of immunology.

The immune system is a masterpiece of complexity, composed of cells and proteins that harmonize exquisitely to respond to threats from outside the body or from within. The immune system is comprised in part of white blood cells, of which some twenty different types are produced mostly in the bone marrow, but also in lymphatic tissue (lymphocytes) and other glands. They circulate freely in the bloodstream until they are called on to fight a local or general infection, or perhaps even a tumor. The appropriate immune cells are attracted to the site by a signaling process called chemotaxis; then their attack is choreographed and conducted by a variety of proteins and peptides that the cells secrete to upregulate and downregulate each other in specific sequences and at precise times. Some immune cells (the B lymphocytes) identify a specific threat and produce specific antibodies against foreign proteins, known as antigens. Other immune cells—such as neutrophils, monocytes, and macrophages—work in concert with various catalytic and regulatory proteins, attacking and digesting the invading antigen, cleaning up the battlefield, and initiating any necessary reconstruction. These cells activate T and B lymphocytes through a signaling process known as costimulation.

Kaplan has an extraordinary ability to express a "sense of the whole" of the interregulating cellular processes. Listening to her describe this cascade of events, one could imagine a field of fireflies in a strong breeze at night, spontaneously coordinating their flickering to spell out a message, word by word. They would need to have an extraordinary system of signals worked out to switch on and off at just the right moments, each in just the right spot.

The immune system is at least that complex. In any particular case of a cascade (or chain reaction) of immunologic responses, there is almost no way to explore or reveal every one of the crucial biochemical steps; it is rather like trying to track the water from a bucket poured down Niagara Falls. Nonetheless, in labs around the world, that is what researchers are trying to do. It is surely a credit to our species that, although we share 99 percent of our DNA sequences with chimpanzees, *Homo sapiens* are trying to untangle our own immune system by way of complicated microscopes, fragile test tubes, and clumsy Latin.

The dynamic complexity of the immune system is what Kaplan emphasizes again and again. "As researchers, we shine a flashlight upon this or that signaling network or response, taking one pathway as our subject, then another, and each time we leave all the other networks in the dark, and we don't notice how they are interacting with our subject. So we generally find what we are looking for—but we never get the whole picture." As she speaks, one can almost picture her trying to reach up and turn on an overhead light, so that she can put away the flashlight.

What Kaplan is pointing out is that, in analyzing the interregulation of immune cells, "we see the world through the lenses we grind." She insists that it is inherently impossible to capture a dynamic, four-dimensional process with two-dimensional snapshots, which is about all that lab testing can provide. You can take a series of snapshots at different moments and from different angles, but it is still just a series of snapshots of a complex dynamic process. Philosophically, Kaplan's view also raises Heisenberg's uncertainty principle from the subatomic level to the subcellular. The only way to observe costimulation is to freeze it in motion, thereby altering it, à la Heisenberg.

Kaplan began testing the effects of thalidomide *in vitro* on immune cells infected with the tuberculosis mycobacteria to find its mechanism of action in fighting symptoms; that is, the collection of chemical pathways whereby the drug affects its remedy. She began to suspect that thalidomide's therapeutic properties might be produced by an action different from either its destructive impact upon the embryo or its potential for nerve damage. Kaplan explains, in the simplest of language: "Normalcy for the immune system is silence, and activation or perturbations take time to quiet down, like ripples in a pond. When the body's defense system is triggered inappropriately by an autoimmune reaction, that silence is broken, and the various types of cells that normally regulate each other in complex networks begin echoing abnormally off each other, continuously disrupting the cascade. Thalidomide seems to have the capacity to restore some of the immune system's signals to silence, allowing it time to re-set itself. That capacity varies significantly with different disorders." But how does thalidomide accomplish that?

Since the parasitic mycobacteria of both tuberculosis and leprosy reside inside macrophages and monocytes, Kaplan gave her attention to the proteins that these types of cells secrete to regulate other cells. Some twenty of these signaling proteins, or cytokines, have been identified that work together like flag signals between cells, if you will, directing a complex naval operation. If those signals are consistently inappropriate, or even badly timed, the battle will be lost. Kaplan notes that "there are only twenty-six letters in the alphabet, but from these, vast libraries can be filled." Among the many cytokines are the interleukins and the interferons; interleukin-2 and interferon-gamma received a great deal of attention in the 1980s, because they could be effectively given to patients as medicine. But the particular cytokine Kaplan concentrated upon was known as Tumor Necrosis Factor-alpha, or TNF-α. This protein had been described originally by another Rockefeller University investigator, Dr. A. Cerami, who was first to detect a factor he called "cachectin," because it was often associated with the wasting condition cachexia, seen in patients with cancer, autoimmune conditions, and tuberculosis.

At the time Cerami was studying macrophages in test tubes, across the street at Memorial Sloane-Kettering Hospital doctors were examining what turned out to be the same molecule in cancer. A protein composed of 157 amino acids, the molecule was given its new name, "tumor necrosis factor-alpha," after these doctors found that, by producing systemic inflammation in laboratory animals, this cytokine actually caused tumors to regress and even to disappear completely. This phenomenon of tumor necrosis was also demonstrated in animals with tumors who were injected with immune factors from a *different* animal with systemic inflammation. In some cases, the presence of one tumor in a body was shown to activate enough of this particular immune factor to suppress the development of other tumors. So cachectin was renamed Tumor Necrosis Factor-alpha, or TNF-α, which many cells of the body produced when under siege. Although this immune factor can inhibit tumor growth under certain circumstances, for those with inflammatory diseases, it only exacerbated the inflammation.

Kaplan found in her studies that leprosy patients suffering from ENL had unusually high levels of TNF-α in their blood and at the site of their lesions; indeed, most of the systemic as well as local tissue damage could be explained by this excessive immune reaction. In 1991 Kaplan published the results of tests showing that thalidomide could diminish TNF-α levels by as much as 70 percent in vitro. Tested again in patients, thalidomide again caused dramatic decreases of TNF-α, along with remarkable clinical improvement in the patients' ENL lesions. This, then, explained the mystery behind Sheskin's unlikely discovery almost thirty years earlier: thalidomide selectively inhibits the production of TNF-α in monocytes and macrophages, thus reducing inflammatory reactions such as ENL.

Kaplan has been traveling to Brazil since 1985 to study leprosy patients who have been treated with thalidomide, and to learn more about the disease she was studying in New York. In addition, she was now assaying the activity of the drug in patients with *M. tuberculae* infection, which causes tuberculosis. But Kaplan had a problem obtaining even the small amounts of thalidomide she needed for testing. Cautiously, she approached the executives at Grünenthal, which still had stockpiles of the compound, and cautiously the company agreed to supply the thalidomide she needed, by special arrangement between Kaplan and the FDA.

What Kaplan did not know at the time was that she could have ordered thalidomide over the phone, right there and then, and picked it up a half hour later, downtown in Greenwich Village.

In 1894, the first seven leprosy patients arrived by barge down the Mississippi at the Gillis W. Long Hansen's Disease Center in Carville, Louisiana. There they received care from an enlightened physician, Dr. Isadore Dyer, who was determined to make this "a place of refuge, not reproach; a place of treatment and research, not detention." Today in the United States, there are about 1,000 patients with leprosy, and some additional 200 people a year contract the disease, probably picking up the mycobacterium from the soil. That is also how the microparasite, M. *leprae*, is acquired by the only other species threatened by leprosy—the armadillo. The armadillo is still indigenous from Louisiana to New Mexico; it is not surprising that there is a higher concentration of human cases there, where the bacterium flourishes.

In the mid-sixties some of the Carville staff, including pharmacologist Dr. Robert Hastings, visited Sheskin in Israel, and saw the extraordinary results of thalidomide on ENL patients. Soon the center began obtaining the drug from the only place it was available: Grünenthal. By the early 1970s the center's director, Dr. Robert Jacobson, had successfully treated numerous patients with thalidomide obtained from Brazil, under special allowances by U.S. Public Health Services, who administered the center. But the supply was erratic, its dosages uncertain, and above all, it was not in a form the body could easily digest; with some of the South American medicine came advice that the pills be chewed, so that they wouldn't simply pass through the patient intact. For a while, according to the center's Dr. Kurt Webster, the Brazilian supply was forwarded to an FDA lab in St. Louis, Missouri, where it was refined and measured carefully before being returned for the patients.

Considering the 92 to 99 percent success rate thalidomide had for ENL, a steady supply of reliably manufactured thalidomide was crucial to many of the center's patients, and, to a few, it was as vital as it was to the first patient Sheskin treated. Without thalidomide, Hastings knew, patients with ENL had to rely upon prednisone (cortisone), which lowered their resistance to tropical infections so severely that their lives were in continual danger. When the erratic supply became life-threatening because of an enormous boom in demand, U.S. Public Health Services allowed the Carville Center to manufacture thalidomide themselves, there on the premises, before it was then double-checked in St. Louis.

That enormous boom in demand was coming from the HIV community.

The human immunodificiency virus, like all other viruses, depends upon living cells. There are two strains of the virus, of which Type 1 is much more common and destructive. When the HIV virus enters the body, it seeks out T-cells, the "commandos" of the immune system, lodges inside them (specifically in T-cells known as CD4+ helper cells), and hijacks their machinery to reproduce itself. There in those T-cells, HIV may remain latent for months or years, until the immune system is severely compromised.

Many of the immune system's cells are kept on a short leash by chemical inhibitors; when unleashed, those cells may be activated inappropriately, as in some phases of HIV, and can wreak havoc on every aspect of every chemical cascade initiated by the immune system as it flails about, looking for some way to stop the virus from spreading through the body and hiding within the cells. External factors may determine the pathways by which the virus proceeds through a particular patient, and therefore what symptoms it produces. In some cases, painful wounds in the mouth called aphthous ulcers may grow and linger for months, as a result of TNF-α being unleashed inappropriately by disruptions in the natural cascade of events. Equally, wasting from AIDS—cachexia—is often fatal in and of itself. With both these symptoms of HIV as well as some others, it turns out that thalidomide works wonders, inhibiting the production of the cytokine TNF-α as it does in ENL and tuberculosis patients with wasting syndrome and, for a time, restoring the patient's immune system to something resembling stasis.

In 1993, working at her lab at the Rockefeller University with several colleagues, Kaplan established that thalidomide suppresses the activation of HIV Type 1 while it is latent. This was extraordinarily promising, with implications that could truly alter the fate of nations, most obviously in sub-Saharan Africa. But scientific discipline did not permit her to cheer yet, given the complex costimulation of the immune system's cells.

Dr. David Stirling is the executive vice president and chief scientific officer of Celgene. Raised in Glasgow, Scotland, Stirling is tall, thin, and unimposing, with a soft voice and a mild Scottish burr to his accent. He did his doctoral studies at the University of

Warwick before coming to the United States to work for the Celanese Corporation as a chemist. In 1986 Celanese spun off Celgene as a small pharmaceutical company in Warren, New Jersey, which began by producing secondary biochemical compounds; that is, substances sold to other manufacturers who synthesized retail pharmaceutical products. The following year, after the new company went public, Stirling began contemplating primary medicines that Celgene might produce directly. He began to consider one of the growing threats related to the AIDS epidemic: new strains of tuberculosis that had become resistant to existing antibiotics. Though rare in most industrialized societies by the 1980s, tuberculosis was on the increase again, taking advantage of the weakened immune system of an AIDS patient and, through improper patient dosing, was developing resistance to many antibiotics.

To learn more about the tuberculosis mycobacterium, Stirling approached a well-known immunologist at Rockefeller University whom he had heard was studying mycobacteria—Dr. Gilla Kaplan. Their first meeting transformed his professional life. She told him about her work with thalidomide, which might well provide a real solution to wasting from tuberculosis, among many other possible conditions. She also told him frankly that—even for her research purposes, much less treating patients—she was unable to find a reliable supply of the drug. The same was true for researchers and patients everywhere, indicating a significant potential market.

Stirling laughs softly as he recalls, in his gentle brogue, how he broke the news to his colleagues that their company should manufacture thalidomide. "When I went back to Celgene and told my associates that we should consider producing thalidomide, the expressions on their faces. . . . " The idea struck him and his colleagues as shocking and absurd, as it does anyone old enough to remember the headlines from the original epidemic.

By this time Kaplan had had extraordinary results in her lab: the drug could actually inhibit replication of HIV-1—at least, in vitro. Could thalidomide be the cure for AIDS? The early indications were ambiguous, and she was engaged in further testing on numerous fronts. As well, more potential uses for the medicine were being discovered by the month, both in theory and in practice. She could not, of course, assess how manufacturing thalidomide could work as a business venture, but scientifically she was convinced that patients with many conditions other than leprosy would benefit from a reliable and consistent source of the drug. Frankly, she was looking for a consistent and reliable supply of the drug for her research.

Stirling had lived in Great Britain fifteen years earlier, during the ordeal of the *Sunday Times*, so he was well aware, in a general way, of just what a controversial idea this was, even if the science was reconfirmed a hundred times. To begin with, why would anyone imagine that the FDA—given the history of thalidomide, Dr. Frances Kelsey, and the reformation of the agency—would even consider a new application for the drug? By this time so much more was known about its dangers than in 1961, when Kelsey had declined to approve it. How could any corporation attract capital to manufacture thalidomide without some assurance from the FDA that they would at least contemplate a well-presented application?

Out of a concern for the patients who could benefit, Kaplan took a gamble. At a dinner party in New York City in about 1994, she had occasion to speak with the commissioner of the FDA, Dr. David Kessler. As she discussed her work with him in a circle of other scientists, she bluntly asked the commissioner how the agency might react to an application for thalidomide's approval. Her question silenced the group.

It was a moment or two before Commissioner Kessler looked at Kaplan and replied, "The agency would consider the application upon its merits, and the evidence of the activities of the drug in the indications to be tested, without prejudice from past events."

Why would the FDA even agree to review such an application? The drug's only proven efficacy was for leprosy patients, whose numbers are small in the United States—and besides, the drug could somehow be supplied to them through compassionate use allowances, without more general approval. What possible reason could there be for the agency to allow any additional risk of birth defects?

The agency had one unassailable reason: to prevent even *more* birth defects, which would surely result from the unregulated traffic of a questionable supply of thalidomide, without any warning to patients. By 1995 the demand for the compound was growing exponentially in the HIV community as its success with more symptoms became known. Kaposi's sarcoma is one example: a form of tumor in the supportive tissues of bone, cartilege, fat, or muscle, producing purple papules that become nodules. It quickly became known—by word of mouth, newsletters, articles, and on the emerging World Wide Web—that even if it was obtained on the black market, for those who were HIV-positive it seemed as if thalidomide was, at last, *the* wonder drug.

The HIV Connection 9

Pawnbroker: "But it is late!"

Dimitri: "For one who comes to a pawnbroker, it is always late."

—Dostoevsky, *The Brothers Karamazov* (1880)

One of the many strange facts in the story of this dark remedy is that in the 1990s a large, collaborative effort was under way between the U.S. government and drug smugglers to make thalidomide available to AIDS patients, a task made more difficult by regulations inspired by . . . thalidomide.

By the early 1980s Grünenthal's patent had expired, leaving the only manufacturers in the world at that time in Brazil and Mexico, although the German company still contributed the drug to researchers and some "special needs" cases.

In the United States, only one small company in Maryland, Andrulis Pharmaceuticals, manufactured a limited amount of the drug each year on a special contract of compassionate use supervised by the FDA, mainly for the Hansen's Disease Center in Carville, Louisiana. But at the same time the black market in various AIDS medicines and other pharmaceuticals had created a new network of drug runners, smuggling medicine from Tijuana to San Diego, for example, for national distribution through buyers' clubs and other "alternative access" channels that had emerged since the beginning of the AIDS crisis in 1981.

Those were the most terrible years of the epidemic, when every day brought new despair to thousands of Americans. The entire culture of San Francisco went into prolonged shock and mourning as thousands upon thousands of young men died brutally without ever understanding the disease that had felled them. Meanwhile, conservative voices across the country, emboldened by the Reagan administration's undisguised contempt for the victims, indulged themselves in an unseemly orgy of sanctimony. These voices, often courting fundamentalist votes, unashamedly suggested that the victims got what was coming to them, and promoted the view that this was just what the gay community should have expected for its unrepentant ways. As a result, any hope that the government would help find a cure or even a vaccine quickly eroded. There simply was no hope.

It isn't known who the first HIV patients were to experiment with thalidomide, but the medical rationale was there even before Gilla Kaplan demonstrated that it inhibited TNF-α secretion, thereby healing mouth ulcers and relieving wasting: because some leprosy patients had tuberculosis, and some tuberculosis patients had AIDS. In any case, during the most brutal, early years of the epidemic, before AZT and the fourteen key elements of today's HIV arsenal, just about any medicine seemed worth trying at some phase of the illness. But few were as effective as thalidomide, at least for aphthous ulcers and wasting.

Matthew Sharp, longtime HIV activist and founder of the Health Alternative Foundation, an information center and drug buyers' club in San Francisco, described the tension with the FDA in the early 1990s. "Because of the FDA's ridiculously slow process of approving new drugs in the 1980s, the HIV community had been smuggling in various pharmaceuticals from Central and South America. When thalidomide started being smuggled in, the FDA really started cracking down on us. The agency had never threatened to close down the drug buyers' clubs before; then all of a sudden they served us with a cease and desist order. But while they were harassing and threatening us, there were other people at the agency working hard to provide thalidomide to some AIDS patients under 'compassionate use' provisions; and gradually they began exploring the possibility of some kind of special approval. There's no question that the HIV community and the drug buyers' clubs around the country played a large role in motivating the FDA to seriously reconsider the drug."

For a new drug to be developed and approved—from test tube to patient—takes on average eight and a half years and costs $359 million, according to FDA estimates. The enormity of the undertaking owes in great part to what was learned from the thalidomide experience. Winning the agency's approval for a new commercial drug is by now a measured procession through a series of stages, beginning with its designation as an Investigational New Drug, or IND, and its submission to the Center for Drug Evaluation and Review (CDER). This is not actually an application for marketing approval, but an exemption from the federal statute prohibiting an unapproved drug from being transported or distributed across state lines. But its main purpose is to document that it is indeed reasonable to proceed with cautious human trials with the drug on the basis of the application's results in animal testing.

A company is expected to test their drug in at least two animal species, for which the FDA provides extensive guidelines. These tests, conducted upon as few animals as possible, must show all relevant data on the absorption of the drug into the bloodstream, as well as its toxicity, breakdown products, and excretion.

Copies of the drug company's application are then distributed between four departments; the overriding review will come from the medical officer, or M.O., who will ultimately determine whether the IND is safe enough to proceed to clinical trials. The M.O. receives three reviews. The chemistry review reports the chemical identity, analysis, and manufacturing control of the drug, basically determining exactly what the patient is receiving, and if there is any apparent threat to human subjects. The pharmacology and toxicology review determines the effects of the drug in animals and whether there is a threat of acute toxicity. It includes a description of the pharmacologic effects and the mechanisms of action of the drug in animals (though the mechanism of action in humans need not be proven for an application to be approved). This review also examines what the effects suggest about human use (as well as information about the absorption, distribution, metabolism, and excretion of the drug). The statistical review analyzes whatever data is provided, and extrapolates what it can.

Following these reviews, CDER has thirty calendar days in which to decide if patients would be at an unacceptable risk, or, importantly, if CDER has the data to make that determination. If CDER raises no red flag, then on day thirty-one after submission of the IND the study may proceed as submitted. That may seem worse than what Frances Kelsey was faced with—except that this clock starts running after the review is done, not sixty days after its submission. Besides, this is only permission to proceed to clinical trials, whereupon the New Drug Application process, or NDA, begins. This process was modified and enhanced in 1985 as part of an attempt to streamline the application procedure; the "NDA Rewrite" of regulations mainly restructured the ways in which data was organized and divided for different panels to review.

The first determination is whether the application is complete; essentially, this allows the medical officer the discretion that was so important to Kelsey with thalidomide. Then, depending on the nature of the drug, the information required may involve a dozen different areas: samples, methods, validation, packaging, and labeling; human pharmacokinetics and bioavailability (i.e., how well the drug enters the system); clinical data and statistics; case reports; and patent information. As well, thanks in part to lessons learned from thalidomide, a Safety Update Report is typically required from the applicant (the drug company), 120 days after the NDA is submitted.

Then human testing begins, first with a small number of healthy volunteers and patients. The protocol for the trial is approved in advance by an established institutional review board (IRB) that carefully monitors the data concerning the subjects' response at different doses, dangers of toxicity, and presence of the drug or its metabolites in different organs. Only then does it advance to the next phase, where the drug is given to patients with the condition for which it has been developed.

This is far more complicated. To begin with, diseases go through different stages with variable symptoms. That is part of the reason why trials are conducted with control groups, patients in comparable conditions who are given a placebo (or nothing at all) instead of the investigational new drug. Obviously, this raises enormous ethical questions—failing to treat subjects who are ill—which the review board considers carefully in designing the protocol for the trials. The solution for this is often "add-on" trials, where all the patients are given the standard therapy, among whom some are also given the new drug.

As Gilla Kaplan put it, "Ultimately, human testing comes down to giving patients a certain drug and later asking how they feel." But how a patient *feels* is highly subjective, and may or may not accurately reflect the medicine's real effects. If the patient believes the drug will work and the doctor consciously or unconsciously conveys the same conviction, it may affect the subjective reactions of the patient. This is why the "gold standard" of drug testing is patient-informed, placebo-controlled, randomized, double-blind trials.

Still, how can the doctor know that the patient wouldn't have improved *without* the new drug? There is no firm answer to this: at best, the physician can only compare the patient's response to the expected progression of the disease. For this reason, the solution lies in testing huge numbers of patients to arrive at statistical rather than anecdotal results.

When all the data from human testing is complete, it is submitted to the FDA's chemists, pharmacologists, and statisticians, and at the end of the process, the medical officer determines whether or not the application demonstrates that the drug is safe and effective for its proposed use in all cases. So, for example, the medical reviewer may seek to reanalyze a drug's effectiveness in a particular patient subpopulation not analyzed in the original submission, or to insist that the sponsor retest effectiveness claims in patient populations determined by the medical officer—pregnant women, for example.

The separate review teams communicate extensively together; the pharmacology reviewer, for example, may work closely with a statistician to evaluate the statistical significance of potential carcinogenic effects in long-term animal studies. Equally, the manufacturing sites are inspected and an array of other final information on labeling, contraindications, and the like is assembled. Then, each reviewer presents conclusions and recommendations. The division director then produces an action letter that provides an approval, approvable, or nonapprovable decision, and a justification for that recommendation.

In 1994, preliminary discussions began between Celgene and the Food and Drug Administration. The agency invited Celgene to demonstrate a clear need for the drug, even though its only *proven* benefits were to leprosy, with just 1,000 U.S. patients (only some of whom have ENL) at Carville's Hansen's Disease Center and similar facilities. That provided very little rationale for even limited FDA approval. Ironically, the genuine but anecdotal successes of the drug with symptoms of HIV could rarely be documented, because the foreign product had never been tested for purity. Although the physicians who were involved in this research could and did begin rigorous clinical testing (i.e., randomized, placebo-controlled, and double-blind) there was often some problem recruiting and keeping test subjects who knew they might be receiving only a placebo, when, through one of the many HIV drug buyers' groups, they could be reasonably certain that they were at least getting the Mexican or Brazilian compound.

But an unusual feature of U.S. drug regulations has great significance to millions of people: the legitimate, "off-label" use of FDA-approved drugs. This means, simply, that once a medication has been approved for treating one condition, doctors may then legally prescribe it *for any other condition*. In short, there is a distinction between "approved drugs" and "approved uses." This feature of the drug laws deserves the scrutiny it has received in recent years: it is not something the public is generally aware of.

To offer a stark example: a doctor who prescribes encainide for a life-threatening heart arrhythmia is following the FDA's "approved use" of the drug, which has been extensively tested for that medical condition. But other doctors are at liberty to prescribe encainide for ventricular premature complexes (VPCs), an off-label use of the drug supported by several peer-reviewed articles. The FDA never approved this use, and when it was eventually tested by the National Institutes of Health (which performs most of the testing requested by the FDA), the death rate for VPCs proved twice as high for patients who received encainide instead of a placebo.

"Off-label use does not represent the gold standard of FDA testing, which calls for randomized, placebo-controlled, double-blind trials," explains Dr. John Swann, Ph.D., the in-house historian (along with Suzanne White Junod) of the Food and Drug Administration. But Congress never intended for the FDA to interfere with the practice of medicine, only to regulate the safety of drugs (after 1938, as a result of sulfanilamide) and their efficacy (after 1962, as a result of thalidomide). It is the responsibility of the state medical boards to oversee the performance of the physicians and how they use the pharmaceutical products that are available.

At the same time, for reasons that are not immediately apparent, the FDA was also mandated to regulate the *promotion* of drugs (as well as medical devices), supervision of which would ordinarily fall to the Federal Communications Commission. But the Kefauver-Harris Act specifically designated this role to prevent the promotion of off-label use. In fact, the FDA holds it as unlawful for drug companies to promote off-label (i.e., unapproved) uses, even by republishing favorable, peer-reviewed journal articles.

What this meant was that the only medical condition for which Celgene's sales force ("detailers," in industry parlance) would be allowed to promote thalidomide was ENL in leprosy patients—largely defeating the whole purpose of the possible approval that was under discussion. Not even Gilla Kaplan's test-tube research could be given out, unless or until the company submitted it to the FDA for verification.

This aspect of the Kefauver-Harris Act has become a serious bone of contention in Washington in the past few years. In 1996 the U.S. Senate proposed a bill (S. 1447) that would have freed the pharmaceutical companies to use a wide variety of methods to promote off-label uses. The agency's deputy commissioner for policy, William B. Schultz, reaffirmed its commitment to upholding this law and cited an observation Senator Kefauver had made in his hearings thirty-four years earlier, regarding the modest claims a pharmaceutical company might make when submitting a new drug to the FDA. In order to expedite approval, he said, "The initial claim would tend to be quite limited. Thereafter, 'the sky would be the limit,' and extreme claims of any kind could be made."

Often the problems of off-label use are related to side effects; the nerve damage caused by thalidomide is an illuminating case in point. As it happens, leprosy causes nerve damage not unlike that produced by the drug; it may be for that reason that these patients almost never experience nerve damage from thalidomide; in fact, some leprosy patients have reported apparent improvement in their peripheral neuritis from the drug. This is not the case for most patients, for whom roughly one in five remain at risk. There is a way to minimize this risk by taking baseline (SNAP) nerve tests before the treatment begins, and then a month or two later; however, this is as yet not fully tested, and may or may not be a reliable indicator. And for the HIV community for whom much of this effort was being made, there is a further complication: among the many drugs these patients take to keep their T-cell counts high and their viral load low, *several* drugs cause peripheral neuropathy.

One important reason why the FDA forbids the promotion of off-label uses is that, as Deputy Schultz put it, "Companies would have no incentive to conduct or fund the necessary scientific research and to present data to FDA to verify the safety and efficacy of those off-label uses. . . . We believe that the risks of allowing drug companies to distribute journal articles and other information about off-label uses far outweigh any benefits."

With thalidomide, then, the question was whether the FDA would ostensibly approve this drug for a few hundred ENL patients, as a means of permitting much broader access to hundreds of thousands of HIV patients, for whom thalidomide's benefits had not been proven and to whom the manufacturer was forbidden to show any relevant studies.

In 1994 Dr. Robert D'Amato and others in Dr. Judah Folkman's lab at the Children's Hospital, Harvard University, published a paper that suggested a dramatic new approach for resolving a variety of malignant tumors without surgery, radiation, or traditional chemotherapy. It was an elegant concept that Folkman, one of the country's most renowned oncologists, had been working on for decades, and with which he had had some success using numerous substances, beginning with cortisone and heparin. Now his solution appeared to lie with thalidomide. This was not owing to its ability to inhibit TNF-α, but rather to one of the drug's unexplored mechanisms of action: antiangiogenesis.

Folkman and his colleagues brought to the effort the important insight that tumors rely on the development of new blood vessels—angiogenesis—for their growth. This is a crucial process for healing an injury, where new vessels are created to deliver nutrients, oxygen, and immune cells and oxygen for reconstruction; that much was well understood. Similarly, it was a process necessary in the development of tumors. It occurred to Folkman that if it were possible to deprive a tumor of its new supply system, perhaps the tumor would die. In a series of tests with rabbit corneas, as well as fertile chicken eggs, he demonstrated that some chemicals could inhibit this process of neovascularization.

In 1991, D'Amato had attended a talk by Folkman on the use of antiangiogenic drugs in cancer therapy. D'Amato was a physician with a Ph.D. in neuropharmacology, and a resident at the Eye and Ear Center at the Children's Hospital in Boston who specialized in ophthalmology. He was fascinated by the concept of antiangiogenic drugs, because the two leading causes of blindness, diabetic retinopathy and macular degeneration, both involve the growth of new blood vessels in the eye. A drug that could stop the growth of new blood vessels might impact those diseases.

Not long after Folkman's talk, D'Amato approached him about doing some work in his lab, where Folkman had an "open projects" board on the wall, listing unassigned projects that the lab hoped to pursue. At the top of the list, going back five years, was the search for an oral antiangiogenic, anticancer drug. At that time all anticancer drugs were injectable, which was significantly less convenient for chronic cancer treatment.

D'Amato reasoned that perhaps such a drug already existed and just hadn't been recognized as such, so he began to search for clues. He was looking for a drug whose side effects included stopping the growth of blood vessels in adult patients—because a key theory of Folkman's work was that, unlike chemotherapy, an antiangiogenic drug would have few side effects in adults, who have very few sites of new vessel development unless they are injured.

Owing to his background in neuropharmacology, the idea of a drug without side effects bothered D'Amato: it is a major tenant of pharmacology that *all* drugs have side effects. So he began a systematic mental search, examining all of the adult body's systems that grow new blood vessels and might therefore exhibit some side effect associated with antiangiogenic therapy. He first considered the skin, hair, and nails. Are new blood vessels required as new hair follicles form? If so, would a side effect of an antiangiogenic drug be hair loss? He continued with the same line of reasoning through each of the various systems.

When he got to the reproductive system, he wondered if sperm production required the formation of new blood vessels and if, therefore, an antiangiogenic drug would cause impotence in males. Then in a flash he realized he was looking at the problem only from a male perspective: abundant new blood vessels grow every twenty-eight days in the female reproductive system! So an antiangiogenic drug should cause problems with a woman's menstrual cycle, and have, as a known side effect, the reduction or termination of a woman's periods (amenorrhea).

He conducted a Medline search on his computer, looking for any drugs that listed amenorrhea as a side effect. There were several hundred that caused amenorrhea by a variety of mechanisms. The list was much too long to plow through experimentally, so he had to think of a different angle. What if the woman was pregnant and not just going through the menstrual cycle? The growing fetus and placenta require the growth of many new blood vessels. What if that growth was stopped? Would the fetus be malformed?

He went back to Medline and searched for all drugs that listed birth defects as a possible side effect. Again, there were too many to test individually. But when D'Amato combined the two lists, only six drugs came up. He stared in disbelief at the name that topped the list: thalidomide.

But his curiosity outweighed his incredulity. D'Amato ordered some thalidomide from Andrulis and set to work testing its antiangiogenicity. Folkman had a ready-made test in his lab: a chicken egg, with a young, developing embryo inside, was broken open into a petri dish, and a pellet containing the test chemical (thalidomide) was placed on the surface of the yolk, before the dish went into an incubator for a few days. Then the pattern of blood vessels on the surface of the yolk was examined: if the test compound was antiangiogenic, a halo would form around the pellet where no blood vessels formed.

On the appointed day for examining the results of the thalidomide test, D'Amato invited Folkman into the lab to witness the dramatic unveiling. But the yolk was perfectly normal, with blood vessels growing everywhere, including around the pellet. Thalidomide had done nothing. Folkman encouraged his young colleague to keep thinking.

What had gone wrong? Where was the flaw in D'Amato's reasoning? Was thalidomide not an antiangiogenic? Should he go on to the next drug on the list? D'Amato was convinced that thalidomide was a good theoretical candidate and he was not ready to give up just yet. He went to the library to see what more he could learn about the drug. There he came across a 1981 paper by Dr. Gary Gordon and others at Johns Hopkins who proposed that thalidomide had to be metabolized by the liver of a sensitive species, such as a rabbit, before it produces its teratogenicity. So D'Amato ran the test again, using one of the metabolic breakdown products of thalidomide. This time it worked. The breakdown product inhibited blood vessel growth on the yolk.

The Folkman lab had another test system for angiogenesis, once the preliminary test had been passed. A small pellet containing FGF-2 (fibroblast growth factor type 2), implanted into a rabbit cornea, stimulated the development of new blood vessels in the cornea. The rabbit could then be fed a test compound to see if the drug would inhibit this FGF-stimulated angiogenesis. Thalidomide fed to the test rabbit made a "huge" difference in the angiogenesis of the cornea. So thalidomide *is* an antiangiogenic drug. After running a number of additional tests, they published their findings.

It was the first suggestion that thalidomide could be used to treat cancer. New doors suddenly began to open.

That same year Dr. Debra Birnkrant from the FDA's Center for Drug Evaluation and Research (CEDR) helped to create the agency's "Working Group on Thalidomide," whose mission was "to ensure consistent practices for the safe use of thalidomide, and to establish a regulatory accounting related to the use of thalidomide, so that the FDA can advise sponsors, pharmaceutical sponsors, as well as individual investigators, and so that we can coordinate the review process for this drug." With that, the agency affirmed what Commissioner Kessler had indicated to Kaplan. Composed of some twenty scientists within the agency, including obstetrician/gynecologists, legal counsel, and other experts, they worked with representatives of the HIV community—especially the drug buyers' groups and ACT-UP, the action and protest organization—and with Celgene, to determine a rational procedure that *might* eventually lead to the availability of thalidomide in the United States.

In a number of meetings held unofficially, and "off-site" from FDA premises, these three unlikely groups together hammered away at the various issues involved, in gatherings that were sometimes raucous and unfocused but which, in the end, provided a model for such cooperation in the future. Brenda Lien, who has spent most of her adult life working in the HIV community, primarily with Project Inform in San Francisco, remembers the climate of collaboration on all sides of the issues as nearly exemplary. "The HIV community took the danger of birth defects very seriously, knowing that, without immense care, their own medical needs could result in terrible future suffering. At the same time, for a regulatory agency, the FDA was making remarkable efforts to actually serve the people."

In May 1997, the National Institute of Allergy and Infectious Diseases (NIAID) conducted a major study with thalidomide as a result of cooperation with the FDA, which is not unusual. The institute's randomized, placebo-controlled, double-blind nationwide study showed a 55 percent rate of complete healing of aphthous ulcers after just four weeks of thalidomide in fifty-seven patients, with a 90 percent rate of improvement in the group. As hopes grew that thalidomide was the best answer yet to HIV and might actually prevent the virus from replicating in patients as Kaplan had shown it did *in vitro*, pressure mounted on the agency to search for a way to make the infamous drug available, safely, to those who needed it.

To that end, in the fall of 1997 an extraordinary conference was held in Bethesda, Maryland. Lasting two days, the event was called an "Open Public Scientific Workshop," under the auspices of the NIH, FDA, and the Centers for Disease Control and Prevention (CDC). Almost fifty speakers addressed the auditorium: they represented researchers, regulators, and victims of the drug; they included oncologists, public health officials, neurologists, leprosy caregivers, AIDS activists, and professors in liability law. Some were there to discuss the benefits of thalidomide, others to warn loudly of its great threat.

One week earlier, at the invitation of the FDA, Celgene had submitted an NDA for thalidomide. By this time the Internet was making it so easy for black-market thalidomide distributors that there were vast, unregulated amounts of the drug traveling all over the country: the potential for disaster was enormous. Along with other Celgene executives, Dr. David Stirling was there; for the company, there really was no telling what to expect. If their application was approved, thalidomide would likely receive designation as an Orphan Drug. Under the 1983 law by the same name, beneficial terms, government research grants, and even some tax breaks were given for drugs that treated medical conditions affecting 200,000 Americans or fewer: thalidomide might even qualify in several different categories. This conference would give the company an opportunity to gauge public sentiment, including the possibility of public outrage, made all the more possible because joining this forum was a well-spoken thalidomide victim, a leader and luminary named Randy Warren.

The moderator of the conference was Stephen C. Groft, the director of rare diseases at the NIH. "I'm sure everyone in this room has suffered through as many sleepless nights as I have during the past year trying to deal with this issue," he began. "It's a hard issue. When we look at the outcomes, not necessarily the benefits, but the bad potential outcomes, it's scary. . . . I guess back in the 1960s, Dr. Frances Kelsey and her colleagues did it right. I think all of us today want to do it right, too."

To open the marathon sessions with a historical perspective, there was Dr. Frances Kelsey herself. Her hair was white by now, and her shoulders a little more hunched than they were when President John F. Kennedy draped a medal upon them thirty-five years earlier, but the same empirical precision, clear of any bias, was still her guiding light. After describing the events of 1961, the woman some had dubbed "a dragon killer" ended her talk emphasizing the beneficial impact the thalidomide experience had had upon the FDA, as well as the monitoring of birth defects and dramatic improvements in the drug regulations of many other nations. "I would say," Kelsey added at the end, "that thalidomide never faded away. I mentioned that we permitted certain trials in cancer to proceed. Some years later, when an application was submitted for leprosy, we felt that was a reasonable use, since there was great need for such a drug in this distressing disease."

Her first-person account of the drug's history provided a crucial bridge between the catastrophe of thalidomide and its newfound promise. By putting her unique imprimatur upon the conference and expressing possibility of approval for the first time, Kelsey opened the gates to the possibility widely anticipated.

Nancy Paller, the mother of an AIDS victim, spoke from the podium about her son, Richard, and how he had almost died from wasting syndrome; thalidomide obtained from Brazil had added years to his life, and she recalled her despair when their illegal source of the drug disappeared. "I was not going to let my son die. I became a mother with a mission. Let me tell you, don't ever get in the way of a mother on a mission." She called her senators, her congressmen, the president, and the FDA. The FDA responded quickly, and soon her son was being administered thalidomide by his local physician. "Because of thalidomide, Richard had a chance to experience so many things that he would not have experienced otherwise. . . . He was able to develop and share the knowledge that love is the only important thing. His last words were, 'Love is all there is.'"

In contrast with this heart–rending appeal came the clear, crisp logic of immunologist Dr. Gilla Kaplan, who brought all of her intellectual gravitas to the podium. "As scientists," she began, "our role, our responsibility is to acquire knowledge and hopefully acquire it in an organized fashion." She described her work, showing how thalidomide inhibits TNF-α, and how "the selectivity of the drug means that you can use it in a more focused way. . . . As you can see, there's a diverse list of diseases or clinical situations which may be potential candidates for immune intervention. Ideally, what we would be looking for is a drug that is better than thalidomide; that is, a bit more active, less toxic, nonteratogenic, that can be targeted for each one of these diseases."

By far the most extraordinary moment during the conference came when Randy Warren addressed the gathering from his wheelchair. Thirty-six years old now, he still had blond hair and a mustache. More important, he still had both feet. He had adamantly refused to let the doctors amputate them, and never wore artificial legs again. Wise and soulful, articulate and emotional, and fully capable of rising above his personal perspective, Warren is a man who stands his ground and will not be swayed, especially when it comes to the chemical he loathes with every twisted bone in his body.

Randy had first learned about the return of thalidomide in 1995, on the eve of one of the first off-site meetings the FDA had held, when a CNN reporter got in touch with him to ask how the victims felt about this development. In fact, he was not entirely surprised: he had always felt, as other thalidomiders did, as if the drug was an evil presence that had never vanished, a ghost, a chimera that stalked its disassembled victims every day of their lives. Didn't he think the victims should have been invited? the reporter asked. Randy couldn't have agreed more.

Soon after, he received a letter of apology from the FDA for their oversight. Randy replied on behalf of the Canadian victims with a simple, direct message: we demand inclusion in this process.

In 1996, 60 *Minutes* broadcast its first story about the revival of thalidomide. By coincidence, the Canadian victims were having a meeting, and a number of them saw the fifteen-minute segment in a bar. Together they watched in horror as a woman swallowed a pill of thalidomide to treat her condition. From that moment the members of the victims' organizations became active in their opposition to unregulated thalidomide, and began developing their collective platform on the issue.

One day an executive with Celgene, Bruce Williams, called and left a message on Randy's voice mail. That had to have been a hard message to leave, and harder to receive. But soon the two men talked, and warily agreed to meet in Toronto, together with Celgene's CEO, Sol Barer. At the meeting, Randy, together with Giselle Cole, a thalidomider from Trinidad who was then the president of Canada's association, spent half of the six hours giving the prospective producers a glimpse into the lives, past and present, of the victims they represented. It was a very grim picture. There was extensive drug addiction, alcoholism, and sexual promiscuity. For many, grinding poverty had never ended. That was why in 1988 Randy had become active in the Canadian victims' efforts to receive higher compensation, as medical care became more expensive and their physical conditions began to deteriorate unnaturally.

Celgene surprised the thalidomiders by agreeing to allow Randy and Giselle real participation in the effort to prevent any pregnancies among thalidomide patients. When the thalidomiders learned that the company was devoting much of its capital and manpower into devising a safe replacement for the drug itself, a degree of real trust began to develop.

So Randy had been invited to Bethesda. He had been at the FDA approval hearing the previous week, where he had been invited as an adviser, but insisted on speaking. And he had become friends with one of his greatest personal heroes—Dr. Frances

Kelsey. In fact, Randy and Kelsey sat together one afternoon, watching the funeral of Princess Diana.

As he approached the stage in his wheelchair, Randy knew there was already an unspoken understanding throughout the large auditorium that this effort would result in some form of distribution. It was also widely expected that, despite the best precautions, deformed babies would eventually be born. So those who were present—mostly dispassionate scientists and government health officials—assumed that the victims would oppose *any* future risk whatsoever that others would be maimed the way he and his constituents were. And if Randy Warren brought out all the world's victims against the approval of thalidomide, no one on this good green earth could fault him.

The hall fell silent as Warren briefly summed up his early life, defined by two doses of thalidomide his mother had taken, thirty-two operations, and his childhood in Shriner's Hospital. He brought with him to the podium the experience of parading day after day for the hospital staff, naked and helpless in his artificial limbs, and the threat of having his feet cut off. But he didn't mention those things. He straightened up his three-foot-three-inch body and sat tall.

Randy's voice is direct and compelling, and his tone often ends with an upward lilt, eh? His words are precise. "As a founder and CEO of the Thalidomide Victims' Association of Canada, I represent the one hundred twenty-five Canadian survivors of the drug that violated our mothers and betrayed their doctors. Many of our mothers and fathers were doctors and pharmacists. I speak for thousands of unborn babies, murdered as they were to grow. I speak for the fifty to sixty percent of us born who died young. There are five thousand of us alive today. . . . We thalidomiders, we're a family. We understand each other without words. We want you to understand us. We are you. We feel. We laugh. We cry. We love, and we are loved.

"We who know pain and uncertain futures extend ourselves, our empathy, to those who suffer from other unjust diseases and conditions where thalidomide might help. Thalidomide, I have heard, takes people out of wheelchairs, and so much more. But please respect and remember the fact that thalidomide put me in my wheelchair. Thalidomide gave many of us shortened arms, or no arms, meaning we can't hug back. We can't comb our own hair. . . .

"We of the Thalidomide Victims' Association of Canada will never accept a world with thalidomide in it. We cannot. But we cannot fight thalidomide. It wins every time. We're forced to prefer regulated thalidomide over illegal thalidomide available on street corners, without warnings. We who know suffering cannot deny quality of life or longer life to others who suffer. We demand mandatory compliance with strict distribution systems. Thalidomide must never be a drug of choice, but always of need or last resort. We demand forced research into new analogs to replace thalidomide, with all of the benefits but without the teratogenic and disconcerting nerve damage side effects. When the new analog or drug is developed, we demand the removal of thalidomide from this planet.

"Thalidomide must always be called thalidomide. *No glory names*." Randy went on to urge the FDA to involve the victims as a reminder to the world of the threat of birth defects, and to anticipate "the day thalidomide is a historical painful memory."

"People pause today during pregnancy before they take a tablet, before they smoke, before they take a drink, knowing that it could possibly affect the fetus. Nobody knew that before us. We are a legacy, and we've given a lot to this world. We would hope for some respect back. . . . Let's bring thalidomide to the forefront so that we can all speak about it and so that we can make those determinations whether the benefits actually outweigh the risks. . . . I want to remind the FDA that we're going to be there. We're going to be watching. . . ."

Thalidomide Approved 10

The potent poison o'ercrows my spirit.
—William Shakespeare, *Hamlet* (1600)

Ten months after the conference, on July 16, 1998, the FDA approved thalidomide for the treatment of ENL in patients with leprosy, under Celgene's trade name, Thalomid™. One can only speculate on what might have happened had Randy Warren and the victims' groups whose interests he represented adamantly resisted approval and rallied public outrage behind their cause. Instead, the victims' pained, reluctant, and highly compassionate acceptance helped to make the drug available—through carefully monitored, restrictive distribution—to relieve suffering and to save lives. Setting a standard of respect and compassion for strangers who were suffering from diseases, Randy Warren's very essence challenges the rest of the world to meet that standard. "We thalidomiders," he said sadly, "are the unwilling watchdogs of this process."

The agency knew, of course, that the vast majority of prescriptions would be written for diseases other than leprosy. By now there were more studies and anecdotal reports of benefits for dozens of medical conditions, because it turns out that to some degree many autoimmune conditions are driven by an excessive TNF-α response: everything from lupus to multiple sclerosis, Behçet's syndrome to inflammatory bowel disease (including Crohn's and ulcerative colitis), has responded well to thalidomide. And new fronts were opening in the war on cancer: it was beginning to seem as if there was no end to the possible uses of thalidomide.

In consultation with the FDA, Celgene created a unique procedure for distribution, involving the active participation of the patient, the doctor, the pharmacist, and the drug company, which they named the System for Thalidomide Education and Prescribing Safety, or STEPS. The company describes this as "the most comprehensive safety and education program in the history of the pharmaceutical industry." It is such a unique, innovative program that Celgene has patented the distribution system.

The monitoring system is administered independently by the Slone Epidemiology Unit at the Boston University School of Public Health. Their director, Dr. Allen Mitchell, is unambiguous about the STEPS program. As he told the *New York Times*, "This is more rigorous than anything I am aware of in the world. Perhaps it is a kind of symmetry, that the agency that made its reputation by prohibiting thalidomide might make it again by developing a system that makes it available with minimum risk."

According to the STEPS protocol, the prescribing physician must register with Celgene before writing the prescription and must personally ensure and warrant that the patient thoroughly understands the danger of birth defects from thalidomide. Since it was not known conclusively at the time whether a male who takes thalidomide might convey it to a woman in his semen (it has since been established that he can), these precautions applied to patients of both genders from the beginning.

The prescribing physician gives the patient a small parcel of information, supplied by Celgene when the doctor registers. First, there is a short videotape, to be viewed at home or in the doctor's office, which explains at length the dangers of birth defects. It ends with a brief plea from Giselle Cole, her beautiful face and eloquent heartfelt words filling the screen. As the image slowly zooms back, it reveals her flipper-like hand, and she gently demands extreme caution against pregnancy with these pills.

Watching the tape is a wrenching experience; moving, but not horrific. It is all the more so for a patient about to take thalidomide, because it is almost surely a last resort, so the patient has already taken a variety of other dangerous, powerful drugs that have failed. Suddenly, this patient with arthritis or a skin disorder or lupus must confront a kind of horror wholly different from his own disease, inevitably, perhaps, forming some kind of subconscious equation between the two. Of course, for people old enough to remember clearly the epidemic forty years ago, it is especially unnerving to suddenly see thalidomide as possibly the difference between life and death, the only life raft in the ocean. But to patients reaching out for the last resort, such misgivings are indulgences.

Under the STEPS program, the patient must learn about and practice two forms of birth control while taking the drug (and for a month before and thereafter, as well), and sign a consent form with the doctor (who may read it to patients who are not literate) to declare an understanding of all the information, guaranteeing that he or she will not share the medication with anyone and will return any unused pills at the end of treatment. This signed document must be given to the registered pharmacist, along with the prescription, in order to obtain the drug. Furthermore, every month the patient must fill out a confidential survey, enumerating every sexual encounter that might result in pregnancy: women of childbearing age must take pregnancy tests monthly, beginning weeks before the treatment, and continuing for a month after. Finally, each 50 mg thalidomide capsule has printed upon it a red circle-and-bar over the profile of a pregnant woman, with the words "DO NOT GET PREGNANT."

Are these precautions sufficient? Clearly, the vast majority of patients will understand the enormous threat to an embryo; even patients who are illiterate will understand, if all the parties adhere firmly to the STEPS regime. But will the husband of every fertile woman taking the drug use effective contraception at all times? What about instances of rape wherein the victim does not accept her right to choose abortion? Time, of course, will tell. Human nature being what it is, though, most everyone involved believes it is a question of *when* another child is born deformed by the drug, not *if*. As Dr. Norman Fost from the University of Wisconsin observed at the workshop conference, "It's important to remember that a zero risk is not achievable. There is no system that will prevent the single birth of a child with phocomelia." So the goal had to be "to find some middle ground that properly balances the interests of future children in not being born with this deformity, and second, the interest of women in getting reasonable access to the drug."

The FDA was mindful of the recent history of Accutane™ (isotretinoin), a powerful and effective medication for acute acne and other conditions, which is also a serious teratogen. Distributed under a mandatory counseling program for patients for more than a decade, follow-up surveys had shown that, when patients are thoroughly informed, they adhere to the conditions set out for use—up to a point. One study found that between 1989 and 1993, there were more than 400 pregnancies reported by women who had been prescribed Accutane™. Of 32 live births, half were malformed. "The agency is watching the situation closely," says one official, guardedly.

But thalidomide has provided its own living deterrents. That is what Randy Warren is saying. "First, look at us thalidomiders, and listen to us; *then* take the drug with great caution, as a last resort, and only until something safer is available." Out of respect for these wishes, Celgene included a photograph of an infant victim in every packet of pills.

There remained the issue of nerve damage, which could be permanent. SNAP nerve tests were recommended for every patient prescribed thalidomide before treatment begins (as a baseline), and monthly during treatment. Unfortunately, little is known about the progression of the neuropathy, and although many patients taking thalidomide experience some degree of tingling in the extremities at doses upward of 200 mgs. a day over a few weeks, it is a very tough call for doctor and patient alike to decide when the threat of permanent nerve damage outweighs the therapeutic value of the drug. But this risk versus benefit decision is different in treating a fatal or incapacitating disease than in choosing a sleeping pill.

The public hue and cry that might have followed approval never came. Around the country, pharmacies and physicians began registering with Celgene when the company launched thalidomide in September 1998, and orders began arriving in big numbers: many patients had been waiting desperately for treatment. By mid-1999, some 15,000 prescriptions had been written.

Finally, HIV patients had a reliable, consistent source for thalidomide. But it was at about this time that their demand started dropping off unexpectedly.

There were problems emerging with the use of the drug for HIV patients, both in vivo and in vitro. It wasn't because of the nerve damage, which many HIV patients were learning to live with from other medications. The problem was thalidomide's sedative side effect. Patients with wasting syndrome were generally so weak that they could barely tolerate the additional effect of sedation from the drug, which sometimes also contributed to their physical and emotional depression.

Use of thalidomide for AIDS also started waning when a variety of new medicines were developed that together fundamentally changed the therapy, as well as the gravity of the threat. HIV had become treatable, and Magic Johnson, among many others, is living proof. There are thousands of people living and working who, a decade ago, may not have expected they would be alive today.

At Children's Hospital, Harvard University, Dr. Judah Folkman was still exploring ways to starve tumors by antiangiogenesis, but by this time he had stopped using thalidomide, though his colleague D'Amato continued with his research. It wasn't that the drug didn't work: it clearly seemed to, and other labs confirmed its antiangiogenic effects in other models. The problem was thalidomide's potential for doing too many things. Besides, by this time Folkman's group had developed and refined two new drugs, angiostatin and endostatin, which had much greater antiangiogenic effects, without the many complications of thalidomide. The announcement of this research caused excitement and made headlines around the world in May 1998. "If you are a mouse and you have cancer," Folkman announced dryly, trying to temper expectations, "then we can help you."

Meanwhile, in her lab at Rockefeller, Dr. Gilla Kaplan was discovering another problem about thalidomide and HIV. This issue wasn't just a side effect.

One of the most promising discoveries she had made in the mid-1990s was that thalidomide stimulated T-cell production very dramatically in test tubes, by somehow triggering the immune system's complex signaling system, making for more and more of these immune "commandos" to fight off infections. She did not know how this came about biochemically, but it seemed extraordinarily promising. When administered to patients, however, the results were much less positive, and in some patients with advanced HIV infection (AIDS), conditions worsened with the drug. At first there was some hope that this was a "paradoxical reaction," which means specifically a response to a drug in which the condition initially worsens before improving. But these patients continued to deteriorate, and were taken off thalidomide.

Kaplan had to search much deeper to find an explanation for the difference between what she saw in the test tube and what patients experienced. The question she focused on was: how does thalidomide alter the immunologic pathway? The answer, she discovered, was subtle.

The cytokine TNF-α is secreted not only by monocytes and macrophages, but also by many other cells of the immune system, and, in fact, throughout the body. TNF-α is also produced by T-cells—and therein lies the problem. Because, although thalidomide inhibits TNF-α production in monocytes and macrophages, it also activates T-cells, thereby increasing T-cell production of TNF-α.

Among the many symptoms in HIV patients, thalidomide remains useful almost exclusively for aphthous and esophageal ulcers. The drug is now one of the lesser weapons in the new armory of medicines. Just eighteen months after the drug was approved in response to the HIV community, only 1 percent of the prescriptions written were for HIV-related conditions. By the late-1990s thalidomide was still what it was the day Mückter and Kunz first produced it: a drug in search of a disease, or at least, a disease more prevalent than leprosy.

But there remained two very promising areas for its efficacy: autoimmune diseases and a variety of cancers. "Hardly anyone with HIV takes thalidomide anymore," Matthew Sharp confirmed. "But one guy who used to get it from our buyers' club still orders it; he's buying it for his father, who has multiple myeloma."

Myeloma is an incurable cancer that usually consists of multiple tumors of the bone marrow, where the body's blood is produced continuously. As the tumors form, bones begin to snap almost spontaneously, if the patient so much as rolls over in bed; even the vertebrae break down. Multiple myeloma strikes about 14,000 Americans each year, and treatment of the disease is, according to the *New England Journal of Medicine*, "notoriously difficult. . . . The five-year survival rate for patients treated with chemotherapy has remained at twenty-nine percent for more than four decades." Most patients are subjected to extensive, high-dose chemotherapy first, after stem cells have been harvested and stored for retransplantation afterward; then their immune system is effectively obliterated to accept the transplant of these cells. But "the high mortality rate," said the *NEJM* in a recent editorial "among patients who undergo allogeneic bone marrow transplantation" (i.e., of their own marrow) "has limited the use of this procedure."

For patients who have already undergone these treatments—and in whom the disease is usually very advanced—thalidomide produced a significant response in 32 percent of the subjects in the study. After forty years without *any* improvement in survival rates, that was worth cheering about. A slew of other reports since has confirmed these findings, and thalidomide is now standard protocol in myeloma treatment at the Mayo Clinic.

The Mechanism of Action 11

The investigator must cultivate a skeptical state of mind toward all hypotheses—especially his own—and be ready to abandon them the moment evidence points the other way.

—Thomas Hunt Morgan, *Experimental Zoology* (1907)

A drug is a chemical that produces some biological response when given to a living organism. Drugs cannot confer new functions within the organism; they can only modify existing functions. Prozac, for example, cannot permanently alter the brain's biochemistry or reroute its biochemical pathways. It merely (and temporarily) shifts the balance of neurotransmitters. As a general rule, drugs exert multiple effects in an organism rather than a single effect. This is a major problem in drug therapy, because many of the effects are unwanted, sometimes dangerous, side effects.

For a drug to function, it must first dissolve into the body's fluids, such as those of the digestive tract. The dissolved drug is then absorbed into the circulation and is metabolized, primarily in the liver, to produce one or more breakdown products of the drug. It is then distributed by the circulation to various tissues of the body, and finally, the drug and its breakdown products leave the body by excretion.

A drug acts by altering functions within the body's various tissues. It can do this in one of three ways—or sometimes a combination. The drug may bind to a specific receptor in a given tissue, much like inserting a key into a lock, either turning the receptor on or off; the drug may bind to an enzyme, shutting down its function; or the drug may have some other effect, such as lodging itself into the cell membrane. These mechanisms are common to all drugs.

During the past forty years, thalidomide has remained an enigma in almost every aspect of its behavior. First, it does not readily dissolve in the body, a major problem for researchers (drugs that can't be put in solution are far more difficult to test). Second, when thalidomide is metabolized by the liver enzymes, it produces well over 100 breakdown products, making it difficult to determine which ones cause thalidomide's various actions. Third, the possible mechanisms of action of the drug in various tissues and various conditions are either completely unknown or are just now beginning to yield to investigation. The mechanism by which thalidomide causes nerve damage may differ from the mechanism that produces its sedative effect, and the drug's activity in treating certain autoimmune conditions is apparently not the same as its curative value for multiple myeloma. The means by which thalidomide affects an embryo may be dissimilar from the action of the drug in adult systems. Most of thalidomide's potential mechanisms of action are not yet understood and because of its many mysteries, thalidomide is one of the most peculiar drugs ever discovered.

The unique benefits that the drug has demonstrated for a growing number of otherwise intractable conditions brings new urgency to the search for its teratogenic mechanism—the means by which it causes birth defects. To date, thalidomide has been used to treat 130 disorders, for some of which it is the only effective means of arresting a patient's progressive deterioration. Discovering thalidomide's various mechanisms may ultimately help lead to new drugs that possess only its therapeutic benefits and none of its dangers.

For nearly thirty years, as an anatomist and embryologist, I have pursued the specific question of how thalidomide causes selective birth defects. That is, why does it cause most (though by no means all) of its deformities in the limbs, ears, and eyes without affecting many other organs, or destroying the entire embryo? Why doesn't the drug cause birth defects in rats and mice? And are the children of thalidomiders—like the offspring of radiation victims—at risk for birth defects? These fundamental questions have haunted thalidomide's victims and driven its researchers for nearly half a century. More than thirty mechanisms have been proposed to explain thalidomide's teratogenic properties alone.

I joined the search for thalidomide's mechanism of action in causing birth defects in 1972 while I was an undergraduate, as part of my general interest in limb development. In 1974, research in thalidomide conducted by Dr. Jay Lash persuaded me to work toward a Ph.D. in anatomy in his lab at the University of Pennsylvania. Lash was a bit eccentric. Tall, slender, and curly-haired, he always wore a pair of clogs and frequently took off for research stints in Helsinki or Woods Hole without much advance warning. His lab had a casual air to it, even though it was filled with serious students. Lash insisted on intellectual freedom, daring his students to try out their ideas and pursue their deepest interests. When I arrived at Penn, I discovered that he had reached a dead end in his search for thalidomide's mechanism. I realized that before the thalidomide mystery could be solved we needed much more general information about the early events of limb development, of which very little was known.

Just two years earlier, Dr. Janet McCredie, a radiologist in the Department of Surgery, University of Sydney, New South Wales, Australia, had advanced a hypothesis of thalidomide's action based on her observation of numerous radiographs of thalidomide victims in England. McCredie observed that the pattern of bone loss in the limbs resembled the pattern of innervation of the limbs. That is, instead of the normal limb bones, there was a specific pattern of bone loss, which she interpreted as resulting from a reduced nerve supply. This led McCredie to propose that just as thalidomide causes nerve damage in adults, in embryos it damages the cells that later produce the sensory nerves of the limb, resulting in failure of the bones to develop normally. Dr. William McBride espoused a similar mechanism. The hypotheses advanced by McCredie and McBride dominated much of the discussion of thalidomide's mechanism during the ensuing decade.

Between 1978 and 1983 I published several letters and papers challenging McCredie's neuropathy hypothesis. In 1983 I coauthored a paper with one of my students, Teresa Strecker, describing an experiment that Teresa had devised to test the hypothesis directly. Strecker's thinking was impeccable: if it were true that thalidomide's damage to an embryo's nerves caused bones to fail to grow in that embryo's limbs, it should follow that those bones depend heavily on nerves for their normal development. The McCredie hypothesis would, therefore, predict that if nerves were somehow prevented from entering the limb, the limb skeleton would not develop normally. So Teresa and I devised a simple experiment to test this hypothesis. We implanted foil barriers into chick embryos, which blocked the nerves from entering the developing limb, and then watched to see what would happen over the next week. What we observed confirmed our original suspicion: the limb skeleton still developed normally, even without help from the nerves. This test suggested McCredie's hypothesis was incomplete at best, and left the door wide open for a more compelling explanation of thalidomide's most devastating mechanism of action.

Many of the chemical processes that occur during the early stages of human development are quite different from those that govern mature life. The study of those processes is called embryology, or more often today, developmental biology, of which limb development is a significant part. To examine those processes involved in forming the limb—and to see how thalidomide interferes with these processes—one must travel deep into the cellular and subcellular world, into the universe of genes and proteins.

Cells are the self-replicating subunits of all organisms. There are a few dozen different cell types, each with its own shape, structure, and composition, but every cell has a nucleus (at some point in its life span), and every nucleus contains DNA (deoxyribonucleic acid), the famous "spiral staircase" of the double helix, which is identical in all the cells of a particular, individual organism. The genes we inherit are encoded in that DNA, and they govern the cells' processes, including their genesis, their functions, and their self-replication.

But if genes are the cells' rulers, proteins are its slaves. They do the real work of the cell. The production of proteins from the information stored in DNA involves two steps, called transcription and translation. A simple analogy illustrates the process: a cook wants a cake recipe that is found only in a reference book in the library. Because this book cannot be checked out from the library, the cook makes a handwritten copy, or transcription, of the recipe. Later, in the kitchen, the information contained in the copied recipe is used to prepare the cake. This process, from recipe to cake, is called translation. The recipe is in a "code" in the recipe book, where, for example, F-L-O-U-R is the code for a white powder made by grinding and bleaching wheat.

In this analogy, DNA is the reference book that contains many recipes for making different proteins; it is too large a molecule to pass from the library (the cell's nucleus) to the kitchen (the ribosomes) where the proteins are prepared; so, just as the book stays in the library, the DNA remains within the nucleus of the cell. The information in the DNA necessary to make a particular protein is copied through transcription, and the copy, called messenger RNA (mRNA), travels out of the nucleus of the cell to find the kitchen (the ribosomes). All the "ingredients" for the "cake" (the final protein) are the twenty amino acids, which are collected by molecules called transfer RNA (tRNA) and brought to the ribosomes for assembly. Proteins may be hundreds of letters (amino acids) long, and are derived from a twenty-letter alphabet of amino acids.

DNA is a bit different: genes may be hundreds to thousands of letters long, using a four-letter alphabet of small molecules called nucleotides: adenine (A), thymine (T), cytosine (C), and guanine (G). Such a small alphabet used in coding is not unusual: both the Morse code (dit, dah) and binary computer code (0, 1) are based on a two-letter alphabet.

As the DNA code is transcribed and then translated to form a protein, a set of three nucleotides, called a codon, is required to code for each amino acid in the protein. This is very similar to the three symbols (dah, dah, dah) of the Morse code coding for the letter "O" in the word FLOUR. By this simple yet amazing feat of nature, a DNA message, written in a four-letter alphabet "codes for" and is translated into a protein, with a twenty-letter alphabet. Some of the three-letter codons in DNA do not code for any amino acid but convey other messages. Some components of the code are much like the capital letter at the beginning of a sentence or the period at the end. For example, one three-letter sequence, TAC, indicates the beginning of a gene, whereas ATC indicates where the gene ends.

But the DNA code is more complicated than the words and sentences on a page. Each sentence (gene) in the DNA code throughout the body is locked down, so that a given gene is not read out in the wrong tissue or at the wrong time. For example, an embryo, fetus, or young child is growing rapidly, and the cells of nearly all the tissues of the body are rapidly dividing to sustain that growth. That rapid cell division is stimulated by a variety of so-called growth factors. Many cells of the body produce these growth factors, which are proteins encoded by the genes during times of rapid development.

When the body reaches its full, adult size, however (which is also genetically influenced), the production of growth factors shuts down. The sentences (genes) that code for these various protein growth factors are locked down so that they can't be turned on again in most adult tissues. But if something goes wrong—if the lock is broken, or opened at the wrong time—in an adult tissue, cells may begin to divide out of control, resulting in a tumor.

The code for the locks on the sentences is based on the same four letters (A, T, C, and G) as that for a gene. It is as if, in the book of DNA, each sentence is preceded by a series of letters that must be decoded before the sentence can be read.

The keys to unlock the coded messages at the beginning of sentences are specific proteins. These "regulator" or "promoter" proteins bind to the "promoter" or "prepromoter" region upstream from the gene (in front of the sentence). The promoter protein binds to the promoter region by "recognizing" certain nucleotide sequences, such as TATA, CCAAT, or GGGCGG. Once the promoter protein binds to this region of the DNA, it allows the gene to be opened and read, thus allowing for transcription and then translation to produce the protein coded for by that gene.

Many of the proteins coded for in the DNA act as enzymes within the cell—submicroscopic fingers that build other molecules within the cell, such as lipids and carbohydrates; or they stimulate thousands of other activities in the cell, including transcription and translation. Most of the proteins made in a given cell are retained within that cell. But some are packaged for export to the cell surface, and a few pass through the cell membrane into the surrounding fluid. TNF-α, for example, is a protein produced inside the cell. It is packaged as part of new cell membrane fragments that form into tiny spheres and are transported to the cell surface where they fuse with the cell's outer membrane. Now, the TNF-α is a cell-surface protein: part is imbedded in the cell membrane and part protrudes from the cell surface. During an inflammatory reaction, an enzyme on the cell surface (one that acts like a tiny pair of scissors) cleaves the TNF-α just where it comes out of the cell. The TNF-α fragment then floats away from the cell and is carried by the circulation until it binds with cell-surface proteins of other cells, stimulating a cascade of events within those cells, including the production of other proteins that contribute to the inflammatory reaction.

By the mid-1980s, researchers had gathered many bits of information about thalidomide's possible mechanism. The drug's general structure had been known for decades, but it wasn't until 1972 that form and function became linked. That year, an excellent paper by N. Åke Jönsson of AB Kabi, Stockholm, revealed that the shape of one portion of the molecule was strikingly similar to that of two nucleotides: adenine (A) and guanine (G). It has since been demonstrated experimentally that thalidomide has a much greater affinity for guanine than for adenine, and it has almost no affinity for the other two nucleotides.

Jönsson also revealed that thalidomide could slide, or intercalate, into the double helix of the DNA. The structure of DNA is like an open spiral staircase with no face plates between the steps. Because of this open structure, thalidomide can slide in between the steps. Although intrigued by thalidomide's ability to intercalate into DNA, Jönsson was at pains to explain its significance in causing birth defects.

But researchers continued to offer hypotheses. Some of these, such as intercalation, looked promising, whereas others, such as the McCredie hypothesis, had been found lacking. Still others were little more than wild guesses.

In 1986 the Teratology Society (the international society for the study of birth defects) held a symposium to commemorate the twenty-fifth anniversary of the thalidomide disaster. It was a great thrill for a young scientist like myself to attend such a symposium. Two of the key figures in the thalidomide story were there, both of whom I knew by name but had never met: Dr. Widukind Lenz and Dr. Frances Kelsey.

As exciting as the symposium was, at the end I felt a pang of disappointment: no one had addressed the issue of mechanism. After all this time, what did we know about how thalidomide worked? How did thalidomide cause limb defects? How could we be certain of preventing another similar disaster if we couldn't answer that question? I had been involved in this research for about eight years by then, mostly testing McCredie's hypothesis. After the symposium, I expressed my disappointment to Dr. Lewis Holmes, president of the Teratology Society, and Dr. Robert Brent, editor of the journal *Teratology*. To my surprise, they invited me to review the proposed mechanisms for a forthcoming issue. Little did I know how much this invitation would affect the course of my career.

From the papers listed in *Index Medicus*, I was able to identify twenty-four proposed mechanisms of action, of which eight could be rejected out of hand, based on more recent data. I concluded in my paper that, as of 1988, we still knew very little about how thalidomide caused birth defects—mainly because we did not know how the limb develops in the first place. Until we learned more about the early events of limb development, the mechanism of action of thalidomide would remain a riddle.

I was determined to help solve some of the mysteries of how limbs develop. Fortunately, I was not starting from scratch. We were beginning to uncover portions of the body's blueprint for growing arms and legs.

The process of limb development involves a cascade of chemical events, in which successions of indispensable proteins and cells act upon one another as catalysts in precise sequence. These interactions are incredibly complex. Some of the major players are the above-mentioned growth factors—proteins that bind to special receptors on the surface of cells. When everything is working properly, these growth factors initiate the cascade of limb development, prompting the mesoderm cells to proliferate and divide. These particular cells later differentiate into bone, muscle, and blood vessels, among other things.

Another critical, early phase in limb formation is angiogenesis: the creation of new, microscopic blood vessels to increase the supply of nutrients to the small protrusions called the limb buds. In the early stage of embryonic development, cells are growing rapidly, and they require a lot of nutrients. When a given tissue is very small, nutrients can diffuse into the tissue from nearby sources. When tissues become larger, however, nutrients must be carried in by blood vessels. Certain proteins known as integrins help to increase the number of blood vessels, thereby providing more nourishment to tissues, such as the limb buds. With the support of growing vessels, the cell mass of the limb bud grows exponentially, expanding and elongating to take on the form of the definitive limbs. Cells within the elongating limb begin to specialize, with the central cells becoming cartilage, giving way to bone.

This much was becoming clear. But there were still critical gaps in our knowledge. Which growth factors, in particular, initiated this cascade? And which integrins stimulated the blood vessels to provide nourishment to the limb buds? And the most crucial question of all: what was thalidomide's role in all this? The drug must interfere in very specific ways without completely destroying the process. How did it do this?

One afternoon in 1992, while I was mowing the lawn I devised an experiment to determine which growth factors were instrumental in limb development. The system worked like this: future limb territories were removed from a chick embryo, placed into a solution containing a chemical whose impact on limbs we wanted to test, and then grafted to another "host" embryo, where the test graft could form an extra limb. Using this system, my students and I examined the effects of various growth factors on the early stages of limb development. We found that two growth factors, in particular—the protein fibroblast growth factor type 2 (FGF-2) and insulinlike growth factor type I (IGF-I)—had a significant stimulatory effect on limb development in our graft system. It didn't take us long to discover that thalidomide could inhibit the stimulatory effect of FGF-2 and IGF-I on the developing limb, causing the graft to fail or become deformed. Others, too, were homing in on the various ways in which thalidomide impeded normal limb growth.

In 1996, Dr. Deither Neubert's group at Freie University in Berlin demonstrated that thalidomide decreases the production of integrins at the cell's surface, which could interfere with their ability to stimulate new blood vessels. And work by Dr. Robert D'Amato in Dr. Judah Folkman's lab at Harvard had also shown that thalidomide inhibits angiogenesis, especially that stimulated by the growth factor FGF-2. The pieces were falling into place.

In 1997, I participated in three symposia on thalidomide. The first two dealt with the pending introduction of thalidomide into the U.S. market, and as the only developmental biologist on the panel, I was asked to address the issue of embryonic risk in reviving the drug. The third symposium, in Cannes, addressed among other things the history and current status of thalidomide, and I was invited to discuss what we now knew about thalidomide's mechanism.

During my presentation at Cannes, I suggested that in this context, if we viewed the cell as a "black box"—we could see the result of a lot of activity, but not the activity itself. I also suggested that the fifteen proposed mechanisms that had not been ruled out could be grouped into a single model, with IGF and FGF stimulating the cell, and the cell reacting by producing integrins, all of which served to encourage the growth of limb buds. If those growth factor–stimulated cells formed the walls of new blood vessels in the limb bud, the surface integrins (especially an integrin called alpha v beta 3) could increase angiogenesis. At least this is how it *should* work.

If thalidomide were introduced at this critical stage, it might interfere anywhere along the pathway—from binding to one or both growth factors or their receptors and preventing their stimulation of the cell to blocking the synthesis of integrins or their transport to the cell surface. The only part of the cell's "black box" that I could "see" was thalidomide's intercalation into DNA. I concluded by stating that, although we still didn't know the exact mechanism, we were rapidly closing in on the answer.

But I was convinced that with a big push, we could discover the answer *now*. My trip home from Cannes was a long one: between taxis, buses, and planes, I had twenty-five hours to mull over what was happening in the "black box." I was convinced that thalidomide's intercalation into DNA must somehow be key to its mechanism, but the biological significance of such a phenomenon was not clear. If thalidomide intercalates too strongly into the DNA, it should cause actual genetic mutations, but that is not what was happening. Furthermore, intercalation did not explain thalidomide's specificity. If thalidomide intercalated into all the DNA of an embryo, it would destroy it completely, instead of causing defects in specific organ systems and leaving most other tissues essentially untouched. An important part of the puzzle was still missing.

The real question was, Where would thalidomide intercalation have the greatest impact? Where would it be able to do the greatest damage? After going over all the possible sites where thalidomide could interrupt the chain of events between growth factor stimulation and angiogenesis, I focused my attention on Neubert's data on thalidomide's inhibition of integrin protein synthesis. After all, if thalidomide intercalated at the promoter sites of the integrin gene, it could disrupt the transcription of that gene. The decline in certain integrins would, in turn, inhibit angiogenesis, which depends on those proteins. Without blood vessels transporting nutrients to rapidly growing embryonic tissues, such as the limbs, their growth would be inhibited and malformations could result.

I had never seen the DNA sequence for the integrin promoter regions, but I was betting that, with the human genome project in full swing, someone else knew it. I reasoned that if the integrin gene promoter region was rich in guanine—the nucleotide that thalidomide seemed to have such a great affinity for—we might have our mechanism. Thalidomide's specificity might be established. As soon as I arrived home, I called one of my graduate students, Brad Fillmore, and together we set out to find the DNA sequence of the integrin promoter. We pored over the published data on integrin promoter sequences to determine what could make the alpha v and beta 3 integrin subunit promoters unique.

To our utter amazement and delight, we discovered that the promoters for both alpha v and beta 3 contain several GGGCGG sequences (called GC boxes). These boxes bind the so-called Sp1 binding proteins, which help initiate transcription. For RNA polymerase (also a protein) to bind to the promoter sequence of the DNA and start transcription, it must form a complex with other proteins, such as Sp1, attached to the DNA.

GC boxes are rare—only about 9 percent of all gene promoter regions are composed of them. This was the major clue I was looking for to unlock the mechanism of thalidomide-induced birth defects: thalidomide intercalates into guanine-rich promoter domains, preventing the integrins from stimulating new blood vessels that would in turn support the growing limb buds.

But to our amazement, what we had discovered was just the beginning. We found that the gene promoter regions of the two growth factors we had investigated, FGF-2 and IGF-I, *also* have several GC binding sites, as do their receptors on the cell surface.

In fact, the genes for at least eight proteins in this growth-stimulating cascade have *exactly* the type of promoters one would predict as being the most highly sensitive to thalidomide. The probability of linking eight GGGCGG, Sp1-dependent promoters in the same pathway is about one in a billion!

The specificity that had been missing from other explanations of thalidomide's mechanism of action leaped out of these data. If thalidomide were to decrease the efficiency of each step in the pathway from growth factors to integrins by only 10 percent, the entire pathway efficiency would be reduced by more than 50 percent. In biochemical terms, that's enough to shut down most pathways. Simply put, limb development depends heavily on a pathway that is highly sensitive to thalidomide.

As a result of our discoveries, we proposed that the IGF-I/FGF-2 → integrin alpha v beta 3 pathway is critical to the angiogenesis and subsequent normal development of some, but not all, embryonic structures. There are multiple angiogenic-stimulating pathways involving different growth factors, receptors, or both, and various embryonic structures differ in their dependence on one or the other of those pathways. Thalidomide affected a number of developing structures in addition to the limb. We believe that the embryonic structures damaged by thalidomide share a common angiogenic pathway, whereas structures that remain unharmed are supplied by blood vessels involving other proteins.

We presented this new, unifying model of thalidomide's mechanism of action at the sixth international limb meeting in Sun Valley, Idaho, in May 1998, and published the model in *Teratology* in April 2000 and *Biochemical Pharmacology* in June 2000. Our hypothesis is just the beginning, however: there is extensive work yet to be done. This is how science goes. Each hypothesis, each new finding, moves the field forward, as we and other researchers work to test it. Time alone will tell if our work has led to a major breakthrough in understanding.

There are at least two other considerations that need to be incorporated into our model. First, there is a fair amount of data, such as the work of Dr. Gary Gordon and his colleagues at Johns Hopkins, who proposed that thalidomide must be metabolized by liver enzymes before it produces its teratogenicity. These data suggest that it may not be the native thalidomide that intercalates into DNA but a slightly modified breakdown product that actually intercalates. Second, Dr. Peter Wells and his group at the University of Toronto have presented some convincing data that oxidative DNA damage is involved in thalidomide's teratogenicity. These data suggest that thalidomide or a metabolite can more effectively intercalate into DNA if the DNA has first suffered some oxidative damage. Neither of these two concepts creates a problem for our model, which can easily accommodate them.

If our hypothesis proves correct, it may help to find a satisfactory replacement for thalidomide. As soon as a truly safe substitute is discovered that can alleviate the suffering of countless adults without shackling unborn children with years of suffering, thalidomide can be retired. When that day comes, as Randy Warren has put it, we who are involved in the many aspects of this drug will gather together for thalidomide's funeral. Until that day, all we have to treat many diseases is thalidomide, with its ever-present threat to the unborn.

One Patient's Account

12

Estragon: I can't go on like this.
Vladimir: That's what you *think. . . .*
—Samuel Beckett,
Waiting for Godot (1954)

Since the FDA granted approval in 1998, doctors in the United States have prescribed thalidomide for more than 130 different conditions. Many of these are autoimmune responses, including multiple sclerosis, lupus (systemic and especially cutaneous or discoid), graft versus host disease (especially in life-saving bone marrow transplants), and inflammatory bowel disease (Crohn's, ulcerative colitis), as well as scleroderma, pyoderma gangrenosum, and numerous rare, intractable skin conditions. As a result, a body of research and documentation on the drug's usefulness is growing apace, and its efficacy at lower doses is being established for these disorders, hopefully diminishing the threat of nerve damage that is believed to be dose related (though other factors, like disease pathology and absorption, also appear to be at work).

It is easy to insist that the dangers of thalidomide outweigh all other considerations and that the mere possibility of further birth defects should keep the drug out of production forever. Much too easy, from the standpoint of patients for whom there is no alternate help whatsoever. Some of these medical conditions have been thoroughly researched, but others have not, and a few are so rare that it is barely possible to study them. For many of these disorders, it simply is not profitable for a manufacturer to shoulder the costs of synthesizing and testing a new drug, knowing how few patients will ever need it. The Orphan Drug Law passed in 1983 did go a long way toward improving the situation for treating rare diseases (diseases with fewer than 200,000 estimated patients), but there are many borderline diseases, and little corporate enthusiasm to address them.

Still, one thinks, why would anyone ever take thalidomide? One good reason is a disease such as pyoderma gangrenosum, an autoimmune condition that is estimated to affect only three people in a million. Although its specific symptoms are unique, the first-person account that follows reflects in a general way the experience of most patients desperate to treat a rare and intractable disease, and for whom FDA approval of thalidomide seemed like nothing less than a matter of life and death.

IT BEGAN WITH A MOSQUITO BITE on my calf that appeared to become infected after a week; perhaps, I thought, from my kitten rubbing against my leg. Day by day the small, ordinary, pink volcano opened wider, and the edges of the open wound turned slightly purple. After two weeks of ordinary first-aid care, I went to my general practitioner, who prescribed an antibiotic.

The appearance of infection diminished, and yet the wound continued to open, slowly, wider and wider. In another month it went from the size of a dime to that of a half dollar, despite continuous treatment. And it was deeper, now, as well, perhaps a quarter inch beneath the skin surface.

It was about this time that the whole thing started to seem mysterious. Surely, it had to be something peculiar about the *mosquito*, not me. The doctor, who had known me for fifteen years, was baffled. At forty-six I was in excellent health, with no medical peculiarities. I had never had a serious disease and hadn't been in a hospital since my tonsils were removed at the age of five. Still, the wound slowly grew larger; it seemed as if its edges were being eaten away continuously. The doctor suspected that it could not heal because of damaged (necrotic) tissue left from the infection, and therefore ordered whirlpool cleansing, followed by "debridement": the removal of that tissue.

Ordinarily that is not an especially painful procedure, and the nurses were quite surprised that, for me, the process that continued for days was akin to deliberate torture. By the time the wound had been "cleaned up," it was almost two inches in diameter. That meant that it was too large now to heal itself, and a skin graft would be necessary.

And so a half-skin graft was taken surgically from beside my hipbone, set down over the wound, and bandaged for ten days. The surgeon's surprise when, he found it had not "taken" was very alarming to me. "Skin grafts always take," he exclaimed, "even if you put them on upside-down. Is it possible you have lupus?"

I certainly didn't think so, because I'd never had any of lupus's terrible symptoms. But I couldn't be sure, and I couldn't make sure, for one simple reason: the health insurance I had at the time covered hospitalization only, not doctors, drugs, or diagnostics, and I was already running out of money for all this care. A few weeks later a plastic surgeon succeeded in coaxing the skin graft to "take," using high-tech bandages that were cruelly expensive, and generally not covered under any insurance plan. Gradually the wound settled down.

Then, within weeks, another "mosquito bite" appeared, on my knee—but by now it was October, and there were no mosquitoes. The wound was spontaneous, like a pimple, and I realized then that the first one had probably been the same. This wound brought to mind an accident I'd had as a kid, age ten, falling off my Flexible Flyer and scraping my knee, right in that very spot. Soon a similar wound appeared on the other knee, and, I recalled, it was in exactly the spot where I had cut my knee when twelve years old, on the edge of a dock, while water-skiing. What was happening to my body? And why were the specters of my childhood injuries returning? Would it eventually cover my whole body, and keep me in a tub like Jean-Paul Marat? My doctor had never seen this condition, so he arranged an appointment with an excellent and popular (i.e., busy) dermatologist.

By the time my appointment came around, I had ten or twelve open wounds, all on my legs, each growing larger slowly, relentlessly, and every one of them was in a spot where I could remember injuring myself when I was younger, like ghosts of my youthful recklessness. At one point I had almost thirty wounds on my arms and legs. It was as if my body had forgotten the trick of healing itself, which I'd watched it do just fine all my life, as we all do, with thousands of nicks, cuts, and scratches. Now, no scabbing occurred. The wounds remained damp, and their bluish borders became thinner and then melted away, widening the wound. To maintain my work routine I had to get up an hour earlier, to clean and rebandage all the wounds. Any new injury to the skin of my legs or hands—a scratch from my playful kitten, for example, or any insect bite—became a long and difficult battle.

The impact was a little like becoming a hemophiliac in midlife, except this was not as life-threatening. Or was it? I still had no idea what it was, what caused it, or what could cure it, if anything. It was chilling, making the rounds from internist to surgeon to emergency room, hearing the same words from different voices. "I've never seen anything like that."

Like many patients with undiagnosed conditions, I felt an immense loneliness. And, it seemed, I had no choice: I would have to become my own doctor, a terrifying prospect in itself. So I set out to learn all about the immune system and autoimmune conditions like lupus—the only clue that I had. I might not be able to heal myself, but at least if anyone ever gave a name to this disease, I would be ready to understand what they were talking about.

Then, wounds began to appear on the backs of my hands as well, and on my elbows. After I shook hands with a colleague at work, there was blood on his fingers.

"Let me tell you right now," I said to the dermatologist when I met her, "If we've discovered a new skin disease, we're naming it after you, not me."

"It's a deal," she replied. With that, Dr. Grace Liang Federman told me the only words I wanted to hear: she felt certain she knew what this was. "Pyoderma gangrenosum," was her diagnosis, and a painful biopsy of one wound confirmed that it matched samples of this condition. Despite its scary name, it has nothing whatsoever to do with gangrene. (That is how I learned that many conditions, cells, and molecules were named by the first scientists who isolated them; very often that name is dead wrong, but hangs around for decades as a testament to misunderstanding.)

"Pyoderma gangrenosum," she explained, "is an extremely rare inflammatory disorder of the immune system; by most accounts, only three out of a million people have it, and so it has been very difficult to study." And no wonder that the other doctors who had seen it did not recognize it. "It is not a contagious disease any more than arthritis or other autoimmune conditions. The etiology (cause) of pyoderma is unknown. About two out of three patients with it have long-term inflammatory bowel disease, like Crohn's." Since I had never had the least symptom of an intestinal problem (and annual checkups had shown nothing), that made me one in a million.

"We can treat pyoderma with cortisone, injected locally or taken orally (prednisone). As well there is a medication called Dapsone, usually used to treat leprosy, that is somewhat effective. We do not know how to cure pyoderma, but in some patients it goes into remission spontaneously." Just the way it first appeared.

At this point I was so relieved to have a diagnosis—to find someone who had a name for what was going on in my body—that the gravity of its implications seemed less threatening for a short time. Besides, the pills did magic. Two days after I started the high doses of prednisone, there was a marked improvement in every wound.

But I was also bouncing off the walls from the prednisone, the experience of which is similar to that of amphetamine. In fact, I cleaned out my garage for the first time in fourteen years, and stacked three cords of firewood, all in a day. Eventually I learned that caffeine should be avoided with prednisone, and that lesson made the speedy effect much more bearable. Besides, the doctor was tapering me off the prednisone quickly, as required.

And so the wounds returned. But now, even with a higher dose of prednisone, the healing was only partial. There were fewer wounds—perhaps from the Dapsone—but those that remained were getting worse.

My life was in turmoil as a result of this strange siege I was under, and the only solace I could find came from learning all I could about this condition, and the misguided immune system that was driving it. "Learn!" That was the advice King Arthur, despondent about Guinevere, received from Merlin the magician, in some recountings of the legend. "Learning is what the mind does best," said the wizened old man, "It will help heal your pain."

Merlin was right. As I studied the various elements of the immune system and the things that go wrong with it, I still felt powerless, but I could also see it was a problem that might be solved. Whether that would happen in my lifetime remained to be seen.

Then gradually the wounds healed, for no apparent reason except perhaps the Dapsone, which I continued to take. For the next eighteen months no spontaneous eruptions appeared, and I avoided accidental wounds by assiduously shunning yardwork and the sharp claws of my cat. Apparently, I was in remission. That ended eighteen months later with a blister on the inner side of my ankle, from an athletic walking shoe.

It took off slowly but relentlessly, and even the highest possible doses of prednisone no longer did a thing. The body becomes as inured to corticosteroids as it does to heroin, for whatever malady it is taken, and patients routinely have to augment their dosage to the brink of toxicity; once they get there, the drug no longer has any effect. Actually, that's not true: there are always the side effects. Prednisone, the most important and powerful anti-inflammatory medicine available, is used to treat scores of conditions, from poison ivy to spinal surgery, and in every case, prolonged use—weeks or months—produces enormously destructive side effects. Most visible is the dramatic weight gain and the signature "moon-faced" appearance that swells up in the third or fourth week. Most pernicious, perhaps, is the damage to the bones, as the medication induces osteoporosis. For someone facing serious surgery—or rather, for their surgeons—prednisone carries many added complications, including a near inability to heal properly. Prednisone indiscriminately downregulates the whole immune system, along with the adrenal gland and a few other useful organs. After about three months the drug causes the hands to tremble badly and produces muscular cramps that linger for hours. By then it is also raising the blood pressure, producing cataracts, damaging internal organs, and possibly increasing the risk of cancers, all the while causing havoc with the nervous system. And at about that same time, it stops working. Still, you have to taper slowly off the drug, wean your system off it, even in the event of emergency surgery.

The wound on my ankle, as ugly as some biblical plague, continued to expand. Everything about pyoderma is slow, but inexorable. By now the wound was three inches in diameter and growing. By midday it was more swollen from walking and it began to hurt and to worsen; by evening it was much worse, even if I'd stayed off it all day.

The next medicine we tried was the scariest. Cyclosporin is given to kidney transplant patients to prevent rejection: it shuts the whole immune system down, leaving the body defenseless against every possible infection, as well as every cancer. For three months I took it, in fear for my life. It did nothing. We also tried the chemotherapy methotrexate. Nothing.

I had been bedridden by the ankle wound some four months by now. That was challenging, since I was living alone just then, which wasn't entirely accidental: my future looked so bleak that I didn't want to take anyone else down, least of all a woman who truly cared for me.

By this time I had full health coverage. At Yale Dermatology in New Haven many of the exotic tests they wanted to do became possible, but they revealed nothing. They probably see as many pyoderma patients there as they do anywhere else in the world, but that's still not a great number: at Philadelphia Hospital they had sent out a nationwide call for patients for twenty years, and saw all of eighteen cases, I'm told. I volunteered to show my wounds at Yale's "grand rounds," when all the doctors and their students visit patients with rare conditions. Lying there naked on a cot, one group in white after another clustered around me, it was as close as I've ever felt to a hooker with the navy in town.

The doctors there seemed fairly convinced that the disease arises from an anomaly in the bone marrow—research in France suggested the same thing. That abnormality is called a benign monoclonal gammopathy, and it is believed to be a dark harbinger of multiple myeloma. They wanted some of my bone marrow, to see how much of it was abnormal, but I declined: I wasn't about to have a new wound pierced in my fragile skin. The doctor at Yale recommended that I be given huge, intravenous doses of cortisone over three to five days at a time, at doses reserved for multiple sclerosis patients. If that failed, he said, I should try to join some clinical trial of thalidomide. "There are only anecdotal reports from Europe with pyoderma," he cautioned, "but they're promising. The drug isn't available in the United States, but the FDA just held a big conference a few months ago in Bethesda, Maryland, and there's talk that it may be approved soon."

Not soon enough. But once again, at massive doses, cortisone (Solumedrol, 3,000 mgs. by IV over three days) did the trick, and at last the ankle wound healed, after keeping me housebound for more than a year. And with a face that belonged in the Macy's Day Parade: I could actually see my cheeks at the bottom of my field of vision.

But within a month there was a new insect bite on the side of my hip. This wound did not grow slowly larger, it grew deeper. Ultimately it looked like the entry wound of a bullet that had lodged an inch beneath the skin. It bled very little, which was additionally eerie. Another three days in the special room of the hospital, hooked up for four hours at a time to an IV, did nothing. A month later we tried again, for five days instead of three, with no results at all. Then a half-dozen satellite wounds started to form around it: untreated, the whole area was sure to collapse into an area perhaps four inches wide and at least half an inch deep.

There were no drugs left to try. I heard about one promising new drug that was a "tumor necrosis factor-alpha receptor fusion protein" that had just been approved for Crohn's, and was thought to inhibit binding of TNF-α, and so I started learning what that meant. I also learned that the first dose of this new medicine was very effective, and that the second could cause life-threatening anaphylaxis.

That left thalidomide as my drug of last resort. It had just been approved by the FDA, and I would be among the first few hundred people in the United States to be prescribed the drug under the STEPS program. I had already studied it thoroughly, and, with my growing understanding of the immune system (after almost five years with pyoderma), it all made sense. By that time, frankly, it seemed clear that, if thalidomide didn't work for me, I would not live very long. Because even if Saint Jude came through for this wound, what about the next scratch from my cat, the next mosquito bite on my ankle? That was too much for a non-Catholic to expect from the saint of hopeless causes. It was not clear *how* pyoderma gangrenosum would kill me, but if my body became one vast, agonizing, and incurable wound, I would probably get hold of a bottle of barbiturates and check out.

I recalled the headlines about birth defects when I was a teenager, and vividly remembered the damaged infants. But I had also been studying the medical literature, learning the language of cells and molecules, macrophages and cytokines. With no formal medical education whatsoever, I was out to solve the mystery of pyoderma gangrenosum. I had learned about TNF-α, and I knew about thalidomide. And I had no options.

I registered with the STEPS program, along with my pharmacist and my dermatologist: she gave me a videotape from the manufacturer Celgene to watch at home, and made sure I understood the necessity for two forms of contraception and the immense danger of causing a pregnancy while I was on the drug. (I suspect that the sex lives of most patients reaching out for a drug of last resort is nothing to write home about.) She also arranged for me to see a neurologist, who performed a painless, needle-free nerve test as a baseline, so that we could detect any early signs of nerve damage. I had read all about that, and the fact that it could be permanent terrified me more than anything else.

I didn't learn anything I didn't already know about the drug from Celgene's videotape, but when thalidomider Giselle Cole appeared, I began to weep—and not for my own sorry lot. Whatever squeamishness I first felt at the sight of her misshapen arm was quickly dispelled by the directness and emotional power with which she delivered her message. She *wanted* me to look at her deformity, that was the reason she went on camera: to show with her own flesh exactly what was at stake. And that was exactly why she had overcome her apparent shyness and her absolute abhorrence for the substance that had devastated her life before it began, to warn those like myself, desperate to get well, of the dreadful consequences *I* could produce through carelessness, once I swallowed a pill, and for up to thirty days after the treatment ended. Yes, thalidomide can be present in semen.

I began with a dose of 50 mgs. and within ten days had raised the dosage to 200 mgs. I always took the pill at night, and given the sedative effect, there was no alternative. It was different from any other sleeping pill I had ever taken.

I found thalidomide bearable, for the most part. But as the dosage rose I began to suffer from an uncommon side effect that made the drug difficult to tolerate: an irregular, arrhythmic heartbeat that was consistently unconstant, from second to second, throughout my whole time on the drug. It wasn't scary, as if a heart attack was starting, but it kept me permanently out of synch with myself, and there was no clear connection between the rhythm of my breathing and the beat of my heart. It woke me in the morning and it was the last thing I felt at night before the medicine pulled me down, deep below the surface of consciousness.

With phony nonchalance I waited to see if the wound would show signs of healing. After two weeks there was no change; prednisone had always worked within days—when it worked. But no change also meant that the wound hadn't grown, and I was clutching for anything that resembled hope.

After five weeks I was pretty sure that the wound was improving, and a week after that I knew: thalidomide could and did heal the wound, slowly and effectively. Clearly, I would need to take it for several more weeks, but it was going to do the job—for this wound, and perhaps for all future wounds.

That was when I felt the nerve damage begin. At first it was like the ordinary numbness and faint tingling one feels when the circulation has been cut off, or if your fingers or toes get cold enough for chilblains. And it was only intermittent: just a few minutes at a time.

Then it grew worse. Like one of those cheap-joke hand-buzzers taped to each finger and to the balls of the feet, it felt as if my extremities were receiving an electric jolt. And now these periods lasted an hour or two.

Should I tell my doctor? Would she take me off the drug immediately? I had watched so many wounds on my legs backslide when I went off prednisone that I dreaded being forbidden my newfound hope. On the other hand the hip wound was definitely stabilizing by now. So I reduced my dosage by myself, and then, with my doctor's agreement, tapered slowly off altogether. The wound continued to heal on its own.

Although I was still feeling pretty beaten up from the last long months, the world began opening up again, and I began to zoom out from my tunnel vision. Dr. Federman published a paper about my case in the *Mayo Clinic Journal.*

I was waiting to see if the next wound to my legs would resolve itself normally when, without warning one afternoon, my intestine ruptured. I collapsed, and when I was taken to the emergency room, a CT scan showed undiagnosed diverticulitis, a ballooning in the wall of the intestine, which had now perforated near the middle of my abdomen. I was in danger of dying from peritonitis much more quickly than I would have been with a burst appendix. A surgeon removed a foot-long section of intestine. And with that, the pyoderma gangrenosum was gone forever, and my health ever since has been 110 percent. The extra 10 percent comes from being glad to be alive.

Six months after the surgery I was discussing a book project with an editor in Boston when she asked if I would have any interest in co–authoring a book about the history and science of thalidomide, together with Dr. Trent Stephens. Well, I'd had a lot of experience that I wanted to put to some use; as Mark Twain put it, anyone who has held a bull by the tail knows four or five things more than someone who hasn't. I was intrigued as an historian by the story of thalidomide, and by then I had learned enough of the science to undertake the job.

Thanks, Merlin.

Thalidomide and Hippocrates

13

We are our own dragons as well as our own heroes,
and we have to rescue ourselves from ourselves.

—Tom Robbins, *Still Life with Woodpecker* (1980)

Thalidomide destroys lives. Thalidomide saves lives.

Everyone involved in the revival of this dark remedy hopes that it will be a short one—five years, perhaps—to be followed by a safe analog, several of which are presently in Phase II testing. These analogs, which Celgene calls IMiDs™ (for immunomodulatory drugs), are being developed only with regard to their potency in inhibiting TNF-α, not to their antiangiogenic properties or any other mechanisms. They have not yet been tested in primates. Of these new compounds, the most successful—known as CDC-501—is proving to be *ten thousand times* more effective than thalidomide. The potential teratogenicity of these new compounds derived from thalidomide will ultimately be tested in a number of animals, including perhaps in vitro tests with the limb buds of chicks. The executives at Celgene are extremely aware of the dangers involved on a day-to-day basis; by now Bruce Williams can almost hear Randy Warren's voice in his head.

But in this interim, until those new analogs are available, thalidomide remains the best and sometimes the only treatment for a variety of cancers, autoimmune conditions, and symptoms of HIV: there are even trials in progress with some genetic diseases. In the first two years since winning FDA approval, Celgene's Thalomid™ has been prescribed to some 20,000 patients outside of clinical tests. For the most part thalidomide is used only as a drug of last resort after a patient has reached late-stage or end-stage disease, when its full efficacy is probably not being achieved. But where the medication has shown success, thalidomide treatment is now being started earlier.

At the Gillis W. Long Hansen's Disease Center in Louisiana, thalidomide continues to be an indispensable cure for ENL, and has dramatically diminished the need for the center itself. The federal government, "in their infinite wisdom," adds Dr. Kurt Webster, sold the center in Carville to the state of Louisiana—partly perhaps because of thalidomide's efficacy—which then moved the remaining patients to a facility in Baton Rouge. There, some forty incapacitated patients remain for permanent care.

The most important application of the drug remains, for the moment, multiple myeloma: tumors of the marrow, which lead to bones breaking while the patient sleeps, skin lesions, progressive weakness, and death. Thalidomide is the first new addition to chemotherapy to be proven effective in forty years. At the University of Arkansas, oncologist Dr. Bart Barlogie, who has conducted the most extensive trials of the drug with myeloma, finds it an indispensable component for treatment.

Barlogie came to treat multiple myeloma for the first time with thalidomide through a circuitous route. As reported by the *New York Times*, in 1997 Manhattan resident Beth Wolmer was desperate to find a cure for her thirty-five-year-old husband, Ira, who had been fighting multiple myeloma for two years. "I was always looking for something new," Wolmer told the *Times*. "I would routinely call scientists in their labs to find out what they were working on."

She eventually learned that Dr. Judah Folkman and his colleague Dr. Robert D'Amato had experimented in chicks and rabbits with various new antiangiogenic treatments, including thalidomide, for other cancers. D'Amato was pursuing the notion that drugs that caused birth defects might do so by preventing the growth of new microvessels—and might work the same way in tumors. Wolmer reached Dr. Folkman late one evening and, she says, when she asked if thalidomide might be worth trying for multiple myeloma, "it was like a lightbulb went off"; Folkman thought the idea added up to a chance worth taking. So when Ira Wolmer's oncologist, Dr. Barlogie, called Folkman, the Harvard specialist recommended trying it.

Barlogie obtained Thalomid™ for trials from Celgene, with the blessing of the FDA. Ira Wolmer died soon after; his disease was too advanced for the drug to help him. But before long another myeloma patient whom Barlogie treated with thalidomide "went into almost a complete remission," the oncologist reported.

In January 1999, Dr. Jayesh Mehta, at that time a member of the Barlogie team in Arkansas, reported on the group's progress at a thalidomide conference at the University of Utah. He discussed the results from eighty-four patients treated over a six-month period. One-third showed a positive response to treatment; but he concluded that the treatment may have been stopped too soon and that even better response may have been seen with longer treatment. Mehta says that, even though the mechanism is unknown, he believes that thalidomide works as an antiangiogenic agent, mainly because other inhibitors of TNF-α did not work in the system. Myeloma exhibits enhanced angiogenesis and extensive neovascularization. It was this behavior of multiple myeloma and the antiangiogenesis work in Folkman's lab that caused the Barlogie group to try thalidomide in the first place.

At a New Orleans conference of the American Society of Clinical Oncology in May 2000, Barlogie presented the results from 169 patients, twice as many as in the study they had published six months earlier. He described thalidomide as the first new drug in three decades to offer such beneficial treatment for multiple myeloma. All these patients had received stem-cell transplants; after other conventional drugs had failed, the response to thalidomide was rapid, and after two years the overall survival was 45 percent, with very few relapses. His "educated guess" was that most of these patients would have died in six months without thalidomide.

At the Mayo Clinic Oncology Center, under the care of Dr. Robert Kyle, thalidomide (200 mg.) is given as a pretreatment, together with dexamethasone, before autologous bone marrow transplant (from and to the same patient) is performed; then, after chemotherapy, thalidomide is often resumed to combat graft versus host disease (which is still a problem with autologous transplants). Also at the Mayo Clinic, Dr. Vincent Rajkumar has begun using thalidomide to treat not only active end-stage and refractory myeloma, but the early, indolent form, soon after diagnosis, and he expects to publish his results in early 2001. In his clinical trials, Rajkumar uses thalidomide as an "add-on" treatment, so as not to deprive any patients of the conventional therapy. By now he and his fellow oncologists at Mayo have overcome their initial skepticism about the drug. Dr. Rajkumar, like all the other oncologists working with this disease, describes thalidomide's effects on myeloma as "real" and "dramatic."

Some seventy-five other cancers have been treated with thalidomide, though often these are desperate, end-stage gambles, tried too late to prove anything. Brain cancers (gliomas) have been especially responsive, though carefully executed trials are still wanting. Placebo-controlled studies pose special ethical problems with fatal diseases, of course. But in June 2000, Celgene and the National Cancer Institute announced a cooperative agreement for extensive testing over the next five years, which will provide far greater resources for the company to perform comprehensive trials around the country.

At Celgene, Dr. David Stirling, chief scientific officer, and Dr. Jerome Zeldis, vice president of medical affairs, believe that for the time being thalidomide itself will be an important adjunctive therapy in cancer treatment and other conditions. Stirling echoes the sentiments of Dr. Gilla Kaplan, who recently joined Celgene's board of directors. "We believe the drug probably hits a signaling target (TNF-α, for example) in many types of cells," says Stirling.

"Under conditions where something has gone wrong, it brings you back to the middle." This is clearly the same "silence of the immune system" to which Kaplan referred.

At this writing, thalidomide is being used experimentally in a variety of solid tumors, including those of the kidney, lung, brain, colon, breast, and skin, as well as sarcomas. Reports are pending.

Thalidomide remains a drug of necessity for the majority of serious, refractory autoimmune conditions, from multiple sclerosis to pyoderma gangrenosum, because almost all of these conditions are driven by TNF-α. In each case, the disease is usually treated with corticosteroids—that is, prednisone, which, for many patients over time, becomes either ineffective or intolerable, as well as downright harmful. If thalidomide just buys these patients time without prednisone, it is already doing them a great benefit.

At the Institute of Arthritis Research in Idaho Falls, Idaho, rheumatologists Dr. Craig Scoville and Dr. James Reading conducted a three-year open-ended trial with thirty-one patients suffering from rheumatoid arthritis. Seventeen patients had to drop out because of "adverse events," including the onset of neuritis. Most of the others showed some improvement. Dr. Scoville is continuing to find the lowest effective dosage and the appropriate medicines to use in conjunction. Of course, no mechanism of action has been determined.

Ulcerative bowel disease, or Crohn's disease (variants of the same condition) is one of the most brutal and common autoimmune diseases, involving open wounds in the intestines and a host of discomforts. According to the Crohn's and Colitis Foundation of America, the disease affects one million Americans equally, throughout all communities. Because Crohn's is driven by excessive TNF-α production in the walls of the intestine, the rationale was there for using thalidomide, but since constipation is a common side effect, trials posed special problems. So Dr. Zeldis of Celgene, in conjunction with the Department of Gastroenterology at Cedars-Sinai Hospital in Los Angeles, ran a trial in 1999 using low-dose thalidomide treatment (maximum 100 mgs. a day), with extremely favorable results. Perhaps most impressive was that, after a month, the patients were tapered off prednisone without relapsing. It may well prove that inflammatory bowel conditions are among the most important uses for thalidomide—and may also establish a lower dose of efficacy in treating other conditions, thereby reducing the threat of nerve damage.

Autoimmune conditions of the skin include cutaneous (or discoid) lupus erythematosus, a common and painful lesion that has eluded treatment, but which responds exceptionally well to thalidomide; so, it seems, does the systemic form of the disease (SLE). This is certainly one of the most promising applications of the drug, as well as a dangerous one: a majority of patients are women who experience onset in their twenties and thirties. There are, as well, a variety of other skin conditions for which thalidomide may be or become indispensable when prednisone no longer works, including Behçet's syndrome, psoriasis, and pyoderma gangrenosum.

The symptoms of HIV infection for which the black market evolved in the early 1990s are now, with two exceptions, better treated with other medications. Aphthous ulcers and Kaposi's sarcoma, however, remain refractory to all other drugs.

In 1996 *60 Minutes* featured a patient who took thalidomide for macular degeneration, raising hopes across the country for those approaching blindness. Recently, during the first major study on about fifty patients with macular degeneration, many dropped out because of side effects and none had significant improvement. A major trial was conducted on fifty patients by Dr. Allen Ho, who led the study at the University of Pennsylvania. "It wasn't promising," Ho said. "There were no miraculous recoveries." None showed significant improvement, and many patients dropped out because of the side effects.

At the Cleveland Clinic in Ohio, Dr. Bruce Cohen is conducting a large test with thalidomide upon a very rare genetic disorder: thirty patients with neurofibromatosis, which used to be known as "elephant man's disease," are enrolled in the study, with as many as 2,000 tumors covering their bodies. This is the first study of its kind, and the results are not in.

That is the picture of thalidomide's clinical applications at the end of 2000. Much more will be known and understood in just the next year or two, owing to the large number of new trials that have begun since FDA approval. And no one doubts, least of all Celgene's officials, that there will be more surprises.

As of this writing, there has not been a single case of fetal exposure to Celgene's Thalomid™.

Dr. William McBride's career, after being celebrated worldwide as the brilliant scientist who first linked thalidomide to birth defects in 1961, is a study in hubris. Ten years after his discovery, McBride was honored by L'Institut de La Vie of Paris, who awarded him the Gold Medal in the famous Hall of Mirrors at the Palace of Versailles. Using his Gold Medal honorarium as seed money, he established his own research foundation in Sydney and sought out charitable support from corporations around the world to study. Needless to say, the celebrity gave Dr. McBride a sharp blade with which to harvest grants and investment capital.

McBride began hearing from parents around the world who had children with birth defects and wanted to find a reason why; he often served as an expert witness in lawsuits. One family in Orlando, Florida, who contacted him in 1978, was suing a drug company for causing their child's deformities: the prescription drug the mother had taken for severe morning sickness was called Bendectin, and the drug company that produced it was Merrell-Dow—"grandchild," if you will, of Richardson-Merrell.

Bendectin, however, was a very different drug from thalidomide, and the research that went into its development and testing was irreproachable. The drug had been used by 33 million women without causing any increase in the rate of birth defects, and it was the only drug for which the FDA had a rating of zero risk during pregnancy. Nonetheless, in part because of McBride's global prestige as an expert in drugs that cause birth defects, the number of court cases mounted. Soon Bendectin was known as a "tortogen-litigen": a drug that causes lawsuits.

As a result of the suits in which McBride testified, Bendectin was taken off the market in 1983, simply to avoid relentless litigation. Pregnant women around the world who had used the drug rushed to have therapeutic abortions. One study concluded that extra hospitalization for women with severe morning sickness, deprived of Bendectin, cost the United States some $73 million between 1983 and 1987. Another, even more sweeping result of McBride's crusade against Bendectin was a Supreme Court judgment in 1993 *(Daubert vs. Merrell Dow Pharmaceuticals)*, ruling that expert witnesses could be excluded if the subject of their testimony did not have the general acceptance of the relevant scientific community; courts and attorneys are still trying to establish some test for "general acceptance."

But how was McBride so certain that Bendectin was responsible? From animal research he conducted at his foundation, came his reply. But his assistants stepped forward to declare that the data McBride had reported in one published study did not correlate with what had happened in the lab. Much of the data reported in the paper were fabricated.

The fact was that McBride had *never* been a research scientist; that was not how he had uncovered the threat of thalidomide. McBride had actually made his discovery, not through studying thalidomide, but by *prescribing* it. Then, when his pregnant patients delivered malformed babies, he made the connection, correctly enough, and reported it to the *Lancet* (either in June 1961, if one believes his story, or six months later). But he was never a research scientist and he lacked a basic knowledge of limb development as well as the commitment to unbiased observation in his work—the commitment embodied by Dr. Gilla Kaplan, for example. Whether racked by guilt for having "caused" birth defects by prescribing thalidomide, driven by overzealous concern for malformed children (as he asserted), or dazzled by his aspirations for a Nobel Prize, McBride lost his grip on the truth. After the undeniable, deliberate misrepresentations in his research were revealed, Dr. William McBride was stripped of his license to practice medicine in Australia in 1994.

McBride, now seventy-four, is the only one who could ever reveal what drove him to overreach. He had been a good and caring obstetrician with patients for decades before his fall; and in 1998, the Australian Medical Registry reinstated his license.

Years ago, Dr. Frances Kelsey moved out of the old prefab structure. Today she is deputy of scientific and medical affairs at the FDA's Office of Compliance in Rockville, Maryland, and although the offices are hardly plush, they do have wall-to-wall carpeting. She remains razor sharp, and her age is as irrelevant to her professional acumen as her white hair and slight stoop. She is the beloved and revered treasure of the Food and Drug Administration. One of her colleagues confirmed that she had received the highest marks possible on a recent job evaluation; another, who asked Dr. Kelsey amiably if she was contemplating retirement, received a cool suggestion about his manners.

In October 2000, Kelsey was inducted into the National Women's Hall of Fame. "I am pleased," she said modestly, "to be honored along with women that I admire, especially those that I personally have worked with in science and medicine."

Professor Widukind Lenz passed away in 1995, and Schulte-Hillen's son, the first thalidomide victim he ever saw, is now a pediatrician himself—one of hundreds of children who beat the odds, and to whom Lenz was a sort of foster father. Randy Warren had also remained very close with the Hamburg pediatrician, as well as Schulte-Hillen, the lawyer who spent almost a decade fighting for the German settlement in which Randy was included.

Randy is sore, in his head and in his heart, but mostly in his thirty-nine-year-old body. His thighs, which resemble, in his own words, two soft watermelons with bones floating loosely within, are becoming more of a concern, and the time and effort required for day-to-day chores is growing. He is proud of the Thalidomide Victim Association of Canada: what they have achieved for their members, and what their members have achieved, both in their own lives and in their courage for supporting the cautious use of thalidomide in the United States until it can be replaced with a safe substitute. He knows that he himself set a towering example of compassion that transcends, morally and intellectually, the twisted, mortal coil he was left with for a body.

But to Randy the drug remains a terrifying specter, his stalker. And the revival of the drug has drained him and his organization emotionally and physically, traveling to speak out about the dangers, polling and debating with all the Canadian members, to achieve consensus and assure solidarity. Meanwhile, his health and that of many of the 5,000 survivors, is very much in question. No one knows what will happen to his bulbous thighs or his curled feet in the future. Bodies were never intended for the kinds of stress that thalidomiders place on their distorted skeletons to cope with the real world. And as degenerative diseases begin to add to their problems—diseases like arthritis, which is more prevalent when bones and cartilege are unnaturally twisted—the future looms as a very difficult time. The possibility that a victim might someday need to be treated with thalidomide is too grotesque for Randy to contemplate, unless he ever really has to; but he allows he is glad that his mother and sister and brother will have this medicine at their disposal if someday it can spare them any suffering.

And he remains frightened that he will one day have to explain to a *new* thalidomide victim why he and his association felt they had to support FDA approval. But if that day comes, the thalidomiders will do whatever they can to help that child, as if one of their own. Still, from time to time he has a nightmare in which the mother of a newborn thalidomider is screaming at him for "selling out." All Randy can answer is that the association is ready to help the child. That sounds hollow to him.

But who *is* responsible if another baby is born with birth defects? The Celgene Corporation, which warned the patient extensively through the STEPS program? The doctor who prescribed it, and counseled the patient on the necessity of two forms of birth control, and obtained the patient's signature to that effect? (It bears repeating that Celgene has found evidence of thalidomide in semen; therefore the danger is for patients of both genders, as the company assumed from the beginning it might be.) Or is it the fault of the government, since the FDA gave its approval for distribution? The NIH, for not funding enough research on birth defects? Beyond that, who will *decide* who is responsible? A medical team? A panel of judges? A jury of peers? The government? The press? These are the questions that the drug will continue to pose until the day it has been superceded and finally eradicated. Till then, it continues to challenge everyone it touches to weigh their moral responsibility carefully.

If the narrative account of thalidomide suggests a parable, then there must be a moral to this story. Through each of its real-world contexts, one underlying motif does emerge, and every individual and every organization touched by thalidomide has confirmed it, almost as a theorem of human behavior: *wherever there is an absence of compassion, individual or collective, a lesser human attribute will fill the vacuum*. Individuals, corporations, and governments who do not set an example of compassion invariably set a very different example. From Grünenthal to Distillers, from Dr. Lenz to Dr. Sheskin, from prosthetics doctors to dismissive politicians, from Kefauver to Harold Evans, from the Law Lords to William McBride, from AIDS activists to myeloma researchers, and from Dr. Kelsey to Randy Warren—all have had their capacity for compassion tested. Some triumphed; many failed.

Everyone who did not place compassion and an absolute commitment to the truth ahead of all other considerations in the use of this medicine loosed a destructive monster. *But the monster was never thalidomide itself*. Because in the hands of healers like Sheskin, Kaplan, Barlogie, Stirling, and Zeldis, to name a few, the drug has saved lives, and prolonged them, and relieved great suffering. It is not at all farfetched to suppose how different this history might have been, had compassion governed the decisionmaking process when thalidomide was invented in 1954.

It is easy to see that, had diligent medical scientists been in charge at Chemie Grünenthal, they would have conducted cautious human trials. Or, if they had conducted such trials carefully, they would have detected the neuropathy. And it is not hard to imagine that they might have done serious testing with fatal diseases, rather than rushing out a sedative. Indeed, some small experiments *had* been conducted with cancer in the 1950s, and Dr.

Frances Kelsey noted that the FDA had allowed Richardson-Merrell to continue testing thalidomide with cancer in 1962, but they soon gave it up. What appears to be one of the most valuable, life-saving applications of thalidomide—multiple myeloma—might have been discovered (if only through unusually broad testing) more than forty years ago. From there other applications might have been found, with few if any victims.

A safe variant of thalidomide is now on the way: the most promising new compound, tested in rabbits, produces no birth defects and has an exponentially greater power to inhibit TNF-α than thalidomide. If human testing, conducted under the most rigorous precautions, confirms its promise, this analog drug will bestow enormous benefits to a wide range of people suffering from certain cancers and autoimmune diseases. Its discovery is a direct result of everything that has been learned from thalidomide. But this is an entirely different drug, with a different name, and, hopefully, the benign potential to heal millions.

NOTES

Chapter 1: An Uncertain Utopia

1 **"Man's New World"** *Life* magazine, October 7, 1957.

4 **"the people of this nation"** Dr. Miles H. Robinson to Harris House Committee on Interstate and Foreign Commerce, *Congressional Records* (hereafter referred to as *CR*), 1962, vol. 108, pt. 14, 87th Congress, Second Session, p. 503.

1 million people used some type of sedative daily The *Sunday Times* insight team, *Suffer the Children: The Story of Thalidomide* (Viking, 1979) p. 42 (hereafter the *Sunday Times* book).

4 billion barbiturates . . . one out of every seven Senator Hubert Humphrey, Senate Hearings Before the Subcommittee on Reorganization and International Organizations of the Committee on Government Operations, *CR*, vol. 108, pt. 14, 87th Congress, Second Session, August 9, 1962, p. 16061.

Prescriptions had nearly quadrupled . . . twentyfold Ibid.

5 **"Will the pharmacologist be able to do better** Aldous Huxley, *Sunday Times* of London, June 10, 1956.

"The ultimate target" *Sunday Times* book, p. 43.

6 **Medical Officer (*Stabsarzt*)** Henning Sjöström and Robert Nilsson, *Thalidomide and the Power of the Drug Companies* (Penguin, 1972) (hereafter referred to as Sjöström/Nilsson), p. 48.

7 **Supracillin** *Sunday Times* book, p. 22.

percentage of the profits Sjöström/Nilsson, p. 50.

"without a qualified leader" Sjöström/Nilsson, p. 49.

8 **Sometime in early 1954** *Sunday Times* book, pp. 12–13.

"It was made from" Dr. Robert Brent on BBC, *Horizon*, "Thalidomide: A Necessary Evil," October 29, 1998.

AZT was first developed Ken Flieger, "Testing Drugs in People," *From Test Tube to Patient*, FDA Special Consumer Report, January 1995.

"righting reflexes" . . . lethal dose . . . CIBA *Sunday Times* book, pp. 15, 17.

9 **Just since 1949 . . . In 1953. . . .** Sjöström/Nilsson, pp. 170, 171.

10 **Russian roulette** Henning Sjöström, ibid., p. 187.

"The Twenty-Three" [and ensuing discussion] Robert Jay Lifton, *Nazi Doctors: Medical Killing and the Psychology of Genocide* (Basic Books, 1986).

11 **"it seemed unnecessary"** Sjöström/Nilsson, p. 190.

12 **Only in Grünenthal's published table** *Sunday Times* book, p. 13.

cross the placenta Sjöström/Nilsson, p. 96.

14 **"jiggle cage"** *Sunday Times* book, p. 16.

Grippex *Sunday Times* book, p. 29.

15 **50,000 "therapeutic circulars"** *Sunday Times* book, p. 30.

The marketing campaign [*passim*], *Sunday Times* book, p. 50.

Distillers Co. Unless otherwise noted, all discussion of Distillers is from *Sunday Times* book, pp. 42–63 *passim*.

Chapter 2: The Epidemic

19 **The glass cracks across** Sylvia Plath, "Thalidomide," 1962.

On Christmas Day [*passim*] Except where otherwise noted, this account of the epidemic in Germany, England, and Australia is drawn from Sjöström/Nilsson, pp. 94–111, *Sunday Times* book, pp. 96–111, Harold Evans, *Good Times/Bad Times* (Athenaeum, 1983), and William McBride, *Killing the Messenger* (1994).

23 **"little note was taken"** Helen B. Taussig, "The Thalidomide Syndrome," *Scientific American*, August 1962, p. 29.

the German measles Dr. Herbert H. Schaumberg, NIH Open Public Workshop Conference, September 9, 1997.

twenty-five compounds Sjöström/Nilsson, p. 171.

24 **a well-disguised "bunker"** Sjöström/Nilsson, p. 47.

31 **Annelise Warren** Interviews with Randy Warren, June 2000.

Chapter 3: The United States in Peril

39 **Vicks . . . 10 million tablets** *Sunday Times* book, p. 71.

Triparanol (Mer/29) Ralph Adam Fine, *The Great Drug Deception: The Shocking Story of Mer/29 and the Folks Who Gave You Thalidomide* (Viking, 1972), and the *Sunday Times* book, pp. 64–69.

40 **Rive Gauche in Georgetown** Author's interviews with Dr. Frances Kelsey, May 2000.

"elixir of sulfanilamide" Sjöström/Nilsson, p. 21.

Drug bill of 1958 Sjöström/Nilsson, p. 31.

41 **a fantastic variety of conditions . . . easier to bear."** *Sunday Times* book, p. 68.

George P. Larrick Some of the following analysis owes to a series of articles by John Lear, science editor of the *Saturday Review*, in 1961–1962, beginning before the thalidomide story broke. These were also reprinted in *CR* September 4, 1962, vol. 108, pt. 14, 87th Congress, Second Session, p. 18499.

42 **This "investigational program"** *Sunday Times* book, p. 71, and Sjöström/Nilsson, pp. 112–114.

44 **Dr. Ray Nulsen** *Sunday Times* book, pp. 81–85.

a fraternity brother of one of the Merrell executives Donald Traci, attorney for U.S. victims, in independent Canadian documentary, "Broken Promises," May 1998.

The newly appointed medical officer The account of Dr. Frances O. Kelsey comes from numerous sources: author's interviews with Dr. Kelsey, Dr. John Swann, FDA historian, and other FDA officials, May 2000; John Lear, "The Feminine Conscience of FDA: Dr. Frances Oldham Kelsey," *Saturday Review*, September 1, 1962, pp. 41–45; Morton Mintz, "Heroine of FDA Keeps Bed Drug Off Market," *Washington Post*, July 15,

46 **Dr. Barbara Moulton . . . the Welch scandal** John Lear, "The Drugmakers and the Government—Who Makes the Decisions?" *Saturday Review*, July 1960.

Senator Kefauver, the long, lean Democrat Kefauver account comes from various sources, including Charles L. Fontenay, *Estes Kefauver: A Biography* (University of Tennessee, 1980) and *CR*. See also notes below for Chapter 6.

50 **an interesting collection of meaningless** *Sunday Times* book, p. 75.

fifty-one exchanges with the company *CR*, Senate, vol. 108, pt. 12, August 8, 1962, pp. 15932–34.

51 **"It was a little late** Author's interviews with Frances Kelsey.

54 **weird differences in wording** Ibid.

Dr. Lenz estimated Dr. Widekund Lenz, speech at 1992 UNITH Congress, Amsterdam.

HEROINE OF FDA Morton Mintz, *Washington Post*, July 15, 1962, p. 1.

55 **President Kennedy presented** Press conference, August 2, 1962.

"standing off the promoters of thalidomide" *CR*, Senate, 87th Congress, vol. 108, pt. 12, August 8, 1962, pp. 15932–34.

"blue babies" For a brief account of Taussig's illustrious career, see "Babies of Blue Babies," *Time* magazine, September 28, 1962.

"The one-third who are so deformed Helen B. Taussig, "The Thalidomide Syndrome," *Scientific American*, August 1962, p. 29.

One evening Dr. Taussig Author's interviews with Dr. Frances Kelsey.

57 **Sherri Finkbine** Sherri Finkbine, *The Lesser of Two Evils*, in A. F. Guttmacher, ed., *The Case for Legalized Abortion Now*, Diablo Books (1967), pp. 15–25, and *Time* magazine, August 3, 1962.

Chapter 4: The Aftermath

61 **"Thalidomide makes people stare,"** Poem by Catherine Purvis in Harvey Teff and Colin R. Munro, *Thalidomide, the Legal Aftermath* (Great Britain: Saxon House, 1976), p. 72.

62 **"Diseases of Medical Progress"** Title of a regular feature in *Clinical Pharmacology and Therapeutics*.

63 **"I have never seen"** *Sunday Times* book, p. 117.

64 **The best estimate** Dr. Widekund Lenz, speech at 1992 UNITH Congress, Amsterdam.

"They held him up for me" [*passim*] Unless otherwise noted, much of this account of the English victims' families comes from the *Sunday Times* book, p. 115 [*passim*].

66 **"one common feature** Harold Evans, *Good Times*, p. 72.

"There was not one family" *Sunday Times* book, p. 121.

"People said they shouldn't be allowed to live" *Sunday Times* book, p. 117.

a couple named Van de Put murdered . . . erupted in cheers *Time* magazine, November 16, 1962.

67 **seized two and a half million pills** Dr. Helen Taussig, "The Evils of Camouflage as Illustrated by Thalidomide," *New England Journal of Medicine* 1819 (1963), p. 92.

Sweden was home to Account of epidemic in Sweden, where Henning Sjöström represented numerous victims, in Sjöström/Nilsson, pp. 131–149.

68 **"cannot be considered"** Sjöström/Nilsson, p. 129.

Randy Warren was nine years old Author's interviews with Randy Warren.

it took another four months . . . "there is every possibility" [*passim*] For epidemic and aftermath in Canada, Sjöström/Nilsson, pp. 136–143.

70 **The first civil case in the U.S. . . . Shirley McCarrick** *Sunday Times* book, pp. 129–134.

71 **West German Ministry of Justice opened the criminal trial** German court case, Sjöström/Nilsson, pp. 207–271, *Sunday Times* book, pp. 122–127.

Chapter 5: Moral Justice and the Press

79 **The battle for the thalidomide victims of Great Britain** Account of the British court battles and the *Sunday Times* of London's role, unless otherwise indicated): author's interviews with Harold Evans, July 2000; *Sunday Times* book, pp. 137–224; Evans, *Good Times;* Teff and Munro, *Thalidomide, the Legal Aftermath.*

80 **"The minister was quite sharp** *Sunday Times* book, p. 143.

81 **"No one can sue the government."** Ibid.

87 **corporations have moral responsibilities** Evans, *Good Times*, p. 75.

to keep me out of jail Ibid., p. 60.

95 **panel of five Law Lords** The appointment process is described in the *Financial Times* of London, April 23, 1999, and the *Sunday Telegraph*, January 9, 2000.

96 **"Looking back, I can hardly believe** Author's interview with Evans, July 2000.

97 **put its judgment in a way** Evans, *Good Times*, p. 83.

the European Commission's report European Commission of Human Rights, *Times Newspapers Ltd. and others against the United Kingdom: report of the European Commission of Human Rights* (Council of Europe, 1977).

98 **spiteful members of a printers' union** Author's interview with Evans.

Chapter 6: Reforming the FDA

101 **12,000 pharmaceutical manufacturing companies . . . [*passim*]** *Newark News*, December 12, 1959. Unless otherwise indicated, all of the figures and quotations in this chapter have been culled from the *Congressional Record*, Senate and House, between 1959 and 1962. There is a detailed account of the Kefauver hearings in Richard Edward McFadyen, "Estes Kefauver and the Drug Industry" (Ph.D. dissertation, Emory University, 1973, UMI Microfilm 73–25), and Richard Edward McFadyen, "Thalidomide in America: A Brush with Tragedy," *Clio Medica* 11, no. 2 (1976): 79–93.

103 **one biographer of Kefauver** Charles L. Fontenay, *Estes Kefauver: A Biography* (University of Tennessee, 1980), p. 379.

The protracted battle Richard Strout, "Two Old Pros," *New Republic*, April 25, 1960.

109 **the "ten commandments" of human research** George J. Annas and Michael A. Grodin, eds., *The Nazi Doctors and the Nuremberg Code: Human Rights in Human Experimentation* (University of Maryland, 1992), p. 186.

Chapter 7: Children's Voices, Strong and Clear

111 **"With everything that was extra** *Sunday Times* book, p. 118.

112 **British thalidomider Kevin Donnellon** BBC *Horizon* documentary, "Thalidomide: A Necessary Evil," aired October 29, 1998.

117 **Tony Melendez** This account is drawn from Tony Melendez with Mel White, *A Gift of Hope: The Tony Melendez Story* (Angelus Media, 1989).

Chapter 8: Lazarus Rises

121 **Jerusalem Hospital for Hansen's Disease** *Jerusalem Post*, June 19, 1997.

122 **Dr. Jacob Sheskin** Account drawn from author's interviews with Dr. Gilla Kaplan and Dr. Robert Hastings, May 2000.

125 **Dr. Gilla Kaplan** Author's interviews with Kaplan and Dr. Victoria Freedman, May 2000.

130 **In 1894, the first seven leprosy patients** Account of the Carville Center from author's interviews with Dr. Kurt Webster and Dr. Robert Hastings; and Dr. Claude Earl Fox, administrator, center's closing ceremony, August 19, 1999.

132 **Dr. David Stirling** Author's interviews with Stirling and Kaplan, May 2000.

Chapter 9: The HIV Connection

138 **Matthew Sharp** Author's interview, May 2000.

139 **For a new drug to be developed . . . Winning FDA approval [*passim*]** *Center for Drug Evaluation and Review Handbook*, 2000, and Jeffrey P. Cohn, "The Beginnings: Laboratory and Animal Studies," *From Test Tube to Patient*, FDA Consumer Special Report, January 1995.

140 **Human testing begins** Ken Flieger, "Testing Drugs in People," ibid.
human testing comes down to Author's interview with Dr. Gilla Kaplan.

141 **In 1994, preliminary discussions began** Author's interviews with Celgene's Dr. Jerome Zeldis, Dr. David Stirling, and Bruce Williams, and with Dr. Debra Birkrant of the FDA's Working Group on Thalidomide, May 2000.

143 **"Off-label use does not represent** Author's interview with FDA historian Dr. John Swann.
the FDA holds it as unlawful But in July 1999 the United States District Court for D.C. disagreed, in a challenge to this FDA tenet, *Washington Legal Foundation v. (FDA Commissioner) Henney*. The FDA is appealing this ruling.

144 **"companies would have no incentive"** Deputy William B. Schultz before the Senate Committee on Labor and Human Resources, February 22, 1996.

145 **In 1994 Dr. Robert D'Amato** *New York Times*, May 3, 1998, p. 1, and Dr. Judah Folkman et al., "Angiogenesis inhibition and tumor regression caused by heparin or a heparin fragment in the presence of cortisone," *Science*, August 19, 1983, 719–725; author's interview with Dr. Robert D'Amato.

published a paper Robert D'Amato, M. S. Loughnan, E. Flynn, and J. Folkman, "Thalidomide Is an Inhibitor of Angiogenesis," *Proceedings of the National Academy of Sciences*, April 26, 1991.

148 **"to ensure consistent practices"** All quotations from this conference are from the transcripts of the Open Public Workshop, September 9–10, 1997, sponsored by the FDA, the NIH, and the Centers for Disease Control.

Brenda Lien Author's interview, June 2000.

151 **Thirty-six years old now** Author interviews with Randy Warren.

Chapter 10: Thalidomide Approved

155 **unwilling watchdog** *New York Times*, July 17, 1998.

156 **more rigorous than anything** Ibid.

it has since been established Author's interview with Dr. Jerome Zeldis.

a small parcel of information Description from Celgene's package inserts.

158 **"It's important to remember"** transcripts of the Open Public Workshop.

Accutane These figures come from Dr. Mitchell at the Open Public Workshop and discussions with FDA officials.

Watching the situation closely Author's interview with Dr. Debra Birnkrant.

159 **problems emerging** Author's interviews with Matthew Sharp, Brenda Lien, and Dr. Gilla Kaplan.

160 **The announcement of this research . . . we can help you."** *New York Times*, May 3, 1998, p. 1.

Meanwhile, in her lab Author's interview with Dr. Kaplan.

161 **notoriously difficult** *New England Journal of Medicine*, editorial, November 1999.

Chapter 11: The Mechanism of Action

163 **The investigator must cultivate** T. H. Morgan, *Experimental Zoology* (Macmillan, 1907), p. 7.

A drug is a chemical Leda McKenry and Evelyn Salerno, *Pharmacology in Nursing* (Mosby, 1992), pp. 42–43

164 **there are well over 100 break-down products** H. Schumacher and others, "The Metabolism of Thalidomide: The Fate of Thalidomide and Some of its Hydrolysis Products in Various Species," *British Journal of Pharmacology* 25, 338–351, 1965.

165 **More than thirty mechanisms have been proposed** Trent Stephens and Bradley Fillmore, "Hypothesis: Thalidomide Embryopathy-Proposed Mechanism of Action," *Teratology*, 61:189–195, March 2000.

research in thalidomide conducted by Dr. Jay Lash James Lash and Lauri Sáxen, "Human teratogenesis: *in vitro* studies on thalidomide-inhibited chondrogenesis," *Developmental Biology* 28, 61–70, 1972.

Just two years earlier, Dr. Janet McCredie Janet McCredie and William McBride, "Some Congenital Abnormalities: Possibly Due to Embryonic Peripheral Neuropathy," *Clinical Radiology* 24, 204–211, 1973.

166 **Between 1978 and 1983 I published a number of letters and papers** Trent Stephens, "Limb Development and Peripheral Nervous System," *Lancet* 1, 282–283, 1978. Trent Stephens, "Role of Neural Crest and Peripheral Nerves in Limb Development." *Lancet* 2,434, 1978, Trent Stephens and Teresa McNulty, "Is Thalidomide Embryopathy a Result of Neuropathy?—No," *Teratology* 23, A63, 1981. Teresa Strecker and Trent Stephens, "Peripheral Nerves Do Not Play a Trophic Role in Limb Skeletal Morphogenesis II" *Teratology* 27, 159–167, 1983, Trent Stephens and Teresa Strecker, "A Critical Review of the McCredie-McBride Hypothesis of Neurotrophic Influence on Limb Morphogenesis." *Teratology* 28, 287–292, 1983.

167 **A simple analogy illustrates the process:** Rod Seeley, Trent Stephens, and Philip Tate, *Anatomy and Physiology*, 5th ed. (McGraw Hill, 2000).

169 **The promoter protein binds to the promoter region by "recognizing" certain nucleotide sequences** Trent Stephens and Bradley Fillmore, "Hypothesis: Thalidomide Embryopathy-Proposed Mechanism of Action," *Teratology* 61:189–195, 2000. Trent Stephens, Carolyn Bunde, and Bradley Fillmore, "Commentary: Mechanism of Action in Thalidomide Teratogenesis." *Biochemical Pharmicology* 59: 1489–1499, 2000.

Many of those proteins coded for in the DNA act as enzymes Seeley, Stephens, and Tate, *Anatomy and Physiology*.

That year, an excellent paper by N. Åke Jönsson N. Åke Jönsson, "Chemical Structure and Teratogenic Properties IV. An Outline of a Chemical Hypothesis for the Teratogenic Action of Thalidomide," *Acta Pharm Suecica* 9: 543–562, 1972.

170 **But researchers continued to offer hypotheses** Trent Stephens, "Proposed Mechanisms of Action in Thalidomide Embryopathy," *Teratology* 38:229–239, 1988.

the Teratology Society (the international society for the study of birth defects) held a symposium in the summer of 1986 The papers presented at this symposium, plus two others, were published in a special issue of *Teratology* 38, 1988.

'I was able to identify twenty-four proposed mechanisms of action. Trent Stephens, "Proposed Mechanisms of Action in Thalidomide Embryopathy," *Teratology* 38: 229–239, 1988.

172 **We found that two growth factors, in particular—the protein fibroblast growth factor type 2 (FGF–2) combined with insulinlike growth factor type I (IGF-I) had a significant stimulatory effect on limb development in our graft system.** Trent Stephens, Carolyn Bunde, and Devyn Smith, "Growth Factors and Limb Initiation." Fifth International Limb Development and Regeneration Conference (abstract), York, England, 1996. Trent Stephens and others, "Thalidomide Inhibits Limb Development Through its Antagonism of IGF-I+FGF–2+heparin." *Teratology* 57: 111, 1998.

Dr. Deither Neubert's group Reinhard Neubert and others, "Down-Regulation of Adhesion Receptors on Cells of Primate Embryos as a Probable Mechanism of the Teratogenic Action of Thalidomide." *Life Sciences* 58: 295–316, 1996.

work by Dr. Robert D'Amato in Dr. Judah Folkman's lab Robert D'Amato and others, "Thalidomide Is an Inhibitor of Angiogenesis." *Proceedings of the National Academy of Science U S A* 91: 4082–4085, 1994.

174 **We pored over the published data on integrin promoter sequences** X. Cao and others, "Cloning of the Promoter for the Avian Integrin Beta 3 Subunit Gene and its Regulation by 1,25-dihydroxyvitamin D3." *Journal of Biological Chemistry* 268: 27371–27380, 1993. J. P. Donahue, N. Sugg, and J. Hawiger, "The Integrin Alpha v Gene: Identification and Characterization of the Promoter Region." Biochim Biophys Acta 1219: 228–232, 1994.

the two growth factors we had investigated, FGF–2 and IGF-I, also have several GC binding sites K. B. Pasumarthi, Y. Jin, and P. A. Cattini, "Cloning of the Rat Fibroblast Growth Factor–2 Promoter Region and its Response to Mitogenic Stimuli in Glioma C6 Cells." *Journal of Neurochemistry* 68: 898–908, 1997. Y. R. Boisclair and others, "Three

Clustered Sp1 Sites Are Required for Efficient Ttranscription of the TATA-less Promoter of the Gene for Insulin-like Growth Factor-binding Protein–2 from the Rat." *Journal of Biological Chemistry* 268: 24892–24901, 1993. A. V. Perez-Castro, J. Wilson, and M. R. Altherr, "Genomic Organization of the Human Fibroblast Growth Factor Receptor 3 (FGFR3) Gene and Comparative Sequence Analysis with the Mouse Fgfr3 Gene." *Genomics* 41: 10–16, 1997. H. Werner and others, "Cloning and Characterization of the Proximal Promoter Region of the Rat Insulin-like Growth Factor I (IGF-I) Receptor Gene." *Biochemical Biophysical Research Communications*. 169: 1021–1027, 1990. D. W. Cooke and others, "Analysis of the Human Type I Insulin-like Growth Factor Receptor Promoter Region." *Biochemical Biophysical Research Communications* 177:1113–1120, 1991. M. Adamo, C. T. Roberts Jr., and D. LeRoith, "How Distinct Are the Insulin and Insulin-like Growth Factor I Signalling Systems?" *Biofactors* 3: 151–157, 1992. M.G. Myers, Jr and others, "Insulin Receptor Substrate–1 Mediates Phosphatidylinositol 3'-kinase and p70S6k Signaling During Insulin, Insulin-like Growth Factor–1, and Interleukin–4 Stimulation." *Journal of Biological Chemistry*. 269: 28783–28789, 1994.

175 **The probability of linking eight GGGCGG** Trent Stephens and Bradley Fillmore, "Hypothesis: Thalidomide Embryopathy-Proposed Mechanism of Action," *Teratology* 61: 189–195, 2000. Trent Stephens, Carolyn Bunde, and Bradley Fillmore, "Commentary: Mechanism of Action in Thalidomide Teratogenesis." *Biochemical Pharmicology* 59: 1489–1499, 2000. These calculations are based on the data from P. Bucher, "Weight Matrix Descriptions of Four Eukaryotic RNA Polymerase II Promoter Elements Derived from 502 Unrelated Promoter Sequences." *Journal of Molecular Biology* 212: 563–578, 1990.

As a result of our discoveries, we proposed Trent Stephens and Bradley Fillmore, "Hypothesis: Thalidomide Embryopathy-Proposed Mechanism of Action," *Teratology* 61: 189–195, 2000. Trent Stephens, Carolyn Bunde, and Bradley Fillmore, "Commentary: Mechanism of Action in Thalidomide Teratogenesis." *Biochemical Pharmicology* 59: 1489–1499, 2000.

176 **Peter Wells and his group** Peter Wells and others, "Oxidative Damage in Chemical Teratogenesis." *Mutation Research* 396: 65–78, 1997. T. Parman, M. J. Wiley, and Peter Wells, "Free Radical-mediated Oxidative DNA Damage in the Mechanism of Thalidomide Teratogenicity." *Nature Medicine* 5: 582–585, 1999.

Chapter 12: One Patient's Account

178 **It began with a mosquito bite** This case history is presented in Dr. Grace Liang Federman, "Recalcitrant Pyoderma Gangrenosum Treated with Thalidomide," *Mayo Clinic Journal* (August 2000).

Chapter 13: Thalidomide and Hippocrates

191 **These analogs** Author's interviews with Celgene's Dr. David Stirling, Dr. Jerome Zeldis, and Bruce Williams.

192 **Gillis W. Long Hansen's Disease Center** Author's interviews with Dr. Kurt Webster and Dr. Robert Hastings.

Barlogie came to treat multiple myeloma *New York Times*, November 1999, p. 1.

193 **In January 1999, Dr. Jayesh Mehta** Author's interview with Dr. Mehta.

At a New Orleans conference *New York Times*, May 22, 2000.

194 **At the Mayo Clinic** Author's interviews with Dr. Vincent Rajkumar and patient Alan Callan.

195 **rheumatologists Dr. Craig Scoville** Author's interview with Dr. Scoville.

Ulcerative bowel disease, or Crohn's Author's interview with Dr. Zeldis.

196 **Dr. Allen Ho** "Thalidomide's Other Side," *Copley News Service*, June 12, 2000.

Neurofibromatosis Author's interview with patient.

197 **Dr. William McBride** Numerous news reports, author's interview (1979), Bill Nicol, *McBride: Behind the Myth* (Australia: ABC Enterprises, Crows Nest); and William McBride, *Killing the Messenger* (1994).

FDA rating of zero risk Nicol, *Behind the Myth*, p. 146.

199 **Years ago, Dr. Frances Kelsey** Author's interview with Dr. Kelsey.

Randy is sore Author's interviews with Randy Warren.

ACKNOWLEDGMENTS

FROM TRENT STEPHENS

I am grateful to Robert D'Amato, M.D., Ph.D. at Harvard Children's Hospital in Boston; Jayesh Mehta, M.D., at the University of Arkansas Myeloma and Transplant Center; and Brad Fillmore, my graduate student and coauthor of my papers.

FROM ROCK BRYNNER

Sincere thanks to all those who have been so helpful and generous with their time. From the medical world: Dr. Gilla Kaplan and Dr. Victoria Freedman, Rockefeller University; Dr. Vincent Rajkumar, Mayo Clinic Oncology Center; Dr. John Shaughnessy, Dr. Joshua Epstein, University of Arkansas Myeloma and Transplant Center; Dr Craig Scoville, Institute of Arthritis Research, Idaho Falls, Idaho; Dr. Kurt Webster and Dr. Robert Hastings, Hansen's Disease Center, Carville, Louisiana; Dr. Grace Liang Federman, Dr. Jay Weiner, and Dr. Joseph Catania in Danbury, Connecticut; in San Francisco, Brenda Lien at Project Inform and Matthew Sharp of the Health Alternative Foundation. At the U.S. Food and Drug Administration, my thanks to Dr. John Swann, Dr. Debra Birnkrant, Ms. Susan Cruzan, and the indomitable Dr. Frances Kelsey; from the Celgene Corporation, Dr. Jerome Zeldis, Dr.

David Stirling, and Bruce Williams. Special thanks to Randy Warren and the Thalidomide Victim's Association of Canada for providing valuable historical material from their archives. And my respect, gratitude, and affection to thalidomiders around the world who have trusted our discretion.

INDEX